T0137041

Tangible Modeling with Open Source GIS

Anna Petrasova • Brendan Harmon • Vaclav Petras
Payam Tabrizian • Helena Mitasova

Tangible Modeling with Open Source GIS

Second Edition

 Springer

Anna Petrasova
Center for Geospatial Analytics
North Carolina State University
Raleigh, NC, USA

Vaclav Petras
Center for Geospatial Analytics
North Carolina State University
Raleigh, NC, USA

Helena Mitasova
Center for Geospatial Analytics
North Carolina State University
Raleigh, NC, USA

Brendan Harmon
Robert Reich School of
Landscape Architecture
Louisiana State University
Baton Rouge, LA, USA

Payam Tabrizian
Center for Geospatial Analytics
North Carolina State University
Raleigh, NC, USA

ISBN 978-3-030-07735-8   ISBN 978-3-319-89303-7   (eBook)
https://doi.org/10.1007/978-3-319-89303-7

© The Author(s) 2015, 2018
Softcover re-print of the Hardcover 2nd edition 2018
This work is subject to copyright. All rights are reserved by the Publisher, whether the whole or part of the material is concerned, specifically the rights of translation, reprinting, reuse of illustrations, recitation, broadcasting, reproduction on microfilms or in any other physical way, and transmission or information storage and retrieval, electronic adaptation, computer software, or by similar or dissimilar methodology now known or hereafter developed.
The use of general descriptive names, registered names, trademarks, service marks, etc. in this publication does not imply, even in the absence of a specific statement, that such names are exempt from the relevant protective laws and regulations and therefore free for general use.
The publisher, the authors and the editors are safe to assume that the advice and information in this book are believed to be true and accurate at the date of publication. Neither the publisher nor the authors or the editors give a warranty, express or implied, with respect to the material contained herein or for any errors or omissions that may have been made. The publisher remains neutral with regard to jurisdictional claims in published maps and institutional affiliations.

Printed on acid-free paper

This Springer imprint is published by the registered company Springer International Publishing AG part of Springer Nature.
The registered company address is: Gewerbestrasse 11, 6330 Cham, Switzerland

Preface

This book introduces Tangible Landscape, an open source tangible user interface for geospatial modeling and visualization powered by GRASS GIS and Blender. With Tangible Landscape a physical and a digital model of a landscape are coupled through a near real-time cycle of 3D scanning, geospatial modeling, and projection. This gives geospatial data an interactive, physical form with which users can intuitively interact. Users can, for example, sculpt new landforms with their bare hands to change the flow of simulated digital water, creating new streams and lakes. They can plant forests by simply placing pieces of colored felt that are immediately rendered as 3D trees. This makes geographic information systems (GIS) far more intuitive and accessible for beginners, empowers geospatial experts, and creates new exciting opportunities for developers.

This second edition introduces a new, faster, more powerful version of Tangible Landscape with new modes of interaction and near real-time 3D rendering. There are new chapters on tangible interaction, 3D rendering and immersion, and landscape design. The updated system configuration chapter describes the new hardware and software for Tangible Landscape. The chapter on tangible interaction describes the new modes of interaction and how they work. The chapter on 3D rendering and immersion describes how Tangible Landscape integrates Blender to automatically generate photorealistic visualizations in near-real time. The landscape design case study concludes the book by demonstrating how many of the topics explored in the preceding chapters can be integrated together.

This book explains how to build a tangible interface for geospatial modeling using free and open source software. It teaches digital fabrication methods for building physical models and introduces 3D modeling and rendering for photorealistically visualizing landscapes. It also provides GIS workflows and Python code snippets for tasks like analyzing topography, simulating surface water flow, analyzing viewsheds, routing trails, and modeling vegetation. The book is meant to help educators, researchers, and professionals in any spatial discipline develop their own applications for classrooms, science communication, scenario planning, or participatory engagement. It should also be a useful resource for learning more about geospatial modeling and visualization with open source software. Latest information about Tangible Landscape project can be found at https://tangible-landscape.github.io.

Raleigh, NC, USA   Anna Petrasova
                                                                       Brendan Harmon
                                                                         Vaclav Petras
                                                                       Payam Tabrizian
                                                                      Helena Mitasova

Contents

Acronyms

2D	Two-dimensional
2.5D	Two-and-a-half-dimensional
3D	Three-dimensional
API	Application programming interface
CAD	Computer-aided design
CAM	Computer-aided manufacturing
CLI	Command line interface
CNC	Computer numerical control
DEM	Digital elevation model
DSM	Digital surface model
GIS	Geographic information system
GNU GPL	GNU General Public License
GRASS	Geographic Resources Analysis Support System
GUI	Graphical user interface
HCI	Human-computer interaction
MDF	Medium density fiberboard
NC	North Carolina
NCSU	North Carolina State University
NURBS	Non-uniform rational b-spline
RGB	Red (R), green (G) and blue (B)
RST	Regularized spline with tension
USB	Universal Serial Bus (device communication standard)
USLE	Universal soil loss equation
SDK	Software development kit
SED	Simplified erosion and deposition
TanGeoMS	The tangible geospatial modeling system
TIN	Triangulated irregular network
TSP	Traveling salesman problem
TUI	Tangible user interface

UNC	University of North Carolina
US	United States (of America)
USGS	United States Geological Survey
USDA	United States Department of Agriculture
UI	User interface

Chapter 1
Introduction

The complex, 3D form of the landscape—the morphology of the terrain, the structure of vegetation, and built form—is shaped by processes like anthropogenic development, erosion by wind and water, gravitational forces, fire, solar irradiation, or the spread of disease. In the spatial sciences GIS are used to computationally model, simulate, and analyze these processes and their impact on the landscape. Similarly in the design professions GIS and CAD programs are used to help study, re-envision, and reshape the built environment. These programs rely on GUIs for visualizing and interacting with data. Understanding and manipulating 3D data using a GUI on a 2D display can be highly unintuitive, constraining how we think and act. Being able to interact more naturally with digital space enhances our spatial thinking, encouraging creativity, analytical exploration, and learning. This is critical for designers as they need to intuitively understand and manipulate information in iterative, experimental processes of creation. It is also important for spatial scientists as they need to observe spatial phenomena and then develop and test hypotheses. With tangible user interfaces (TUIs) like Tangible Landscape one can work intuitively by hand with all the benefits of computational modeling and analysis. This chapter discusses the evolution of tangible user interfaces and the development of Tangible Landscape. This chapter also describes the organization of this book.

1.1 Tangible User Interfaces

Inspired by prototypes like Durrell Bishop's Marble Answering Machine (Poynor 1995) and concepts like Fitzmaurice et al.'s Graspable User Interface (1995), Ishii and Ullmer (1997) proposed a new paradigm of human-computer interaction—tangible user interfaces (TUIs). They envisioned that TUIs could make computing more natural and intuitive by coupling digital bits with physical objects as Tangible

© The Author(s) 2018
A. Petrasova et al., *Tangible Modeling with Open Source GIS*,
https://doi.org/10.1007/978-3-319-89303-7_1

Fig. 1.1 Tangible
Landscape: a real-time cycle
of 3D scanning, geospatial
computation and 3D
modeling, and projection and
3D rendering

Bits. In their vision Tangible Bits bridge the physical and digital, affording more
manual dexterity and kinesthetic intelligence and situating computing in physical
space and social context (Ishii and Ullmer 1997; Dourish 2001). Recently, the
development of TUIs has gained momentum thanks to new developments in 3D
technologies such as 3D scanning and 3D printing.

We can easily, intuitively understand and manipulate space physically, but our
understanding is largely qualitative. We can also precisely and quantitatively model
and analyze space computationally, but this tends to be less intuitive and requires
more experience. Intuition allows us to perceive, think, and act in rapid succession;
it allows us to creatively brainstorm and express new ideas. TUIs like Tangible
Landscape (Fig. 1.1) aim to make the use of computers more intuitive combining
the advantages of physicality and computation.

Spatial thinking—'the mental processes of representing, analyzing, and drawing inferences from spatial relations' (Uttal et al. 2013)—is used pervasively in everyday life for tasks such as recognizing things, manipulating things, interacting with others, and way-finding. Higher dimensional spatial thinking—thinking about form, volume, and processes unfolding in time—plays an important role in science, technology, engineering, the arts, and math. Three-dimensional (3D) spatial thinking is used in disciplines such as geology to understand the structure of the earth, ecology to understand the structure of ecosystems, civil engineering to shape landscapes, architecture to design buildings, urban planning to model cities, and the arts to shape sculpture.

Physical models are used to represent landscapes intuitively. With a physical model we can not only see its volume and depth just as we would perceive space in a real landscape, but also feel it by running our hands over the modeled terrain. We can shape physical models intuitively—for example we can sculpt landforms by hand, place models of buildings, or draw directly on the terrain. With a physical model, however, we are constrained to a single scale, simple measurements, and largely qualitative impressions.

Many spatial tasks can be performed computationally enabling users to efficiently store, model, and analyze large sets of spatial data and solve complex spatiotemporal problems. In engineering, design, and the arts computer-aided design (CAD) and 3D modeling software are used to interactively, computationally model, analyze, and animate complex 3D forms. In scientific computing multidimensional spatial patterns and processes can be mathematically modeled, simulated, and optimized using geographic information systems (GIS), geospatial programming, and spatial statistics. GIS can be used to quantitatively model, analyze, simulate, and visualize complex spatial and temporal phenomena—computationally enhancing users' understanding of space. With extensive libraries for point cloud processing, 3D vector modeling, and surface and volumetric modeling and analysis, GIS are powerful tools for studying 3D space.

GIS, however, can be unintuitive, challenging to use, and creatively constraining due to the complexity of the software, the complex workflows, and the limited modes of interaction and visualization (Ratti et al. 2004a). Unintuitive interactions with GIS can frustrate users, constrain how they think about space, and add new cognitive burdens that require highly developed spatial skills and reasoning to overcome. The paradigmatic modes for interacting with GIS today—command line interfaces (CLI) and graphical user interfaces (GUI)—require physical input into devices like keyboards, mice, digitizing pens, and touch screens, but output data visually as text or graphics. Theoretically this disconnect between intention, action, and feedback makes graphical interaction unintuitive (Dourish 2001; Ishii 2008b). Since users can only think about space visually with GUIs, they need sophisticated spatial abilities like mental rotation (Shepard and Metzler 1971; Just and Carpenter 1985) to parse and understand, much less to manipulate 3D space.

In embodied cognition higher cognitive processes are grounded in, built upon, and mediated by bodily experiences such as kinesthetic perception and action (Anderson 2008). Tangible interfaces—interfaces that couple physical and digital data (Dourish 2001)—are designed to enable embodied interaction by physically manifesting digital data so that users can cognitively grasp and absorb it, thinking with it rather than about it (Kirsh 2013). Embodied interaction should be highly intuitive—drawing on existing motor schemas and seamlessly connecting intention, action, and feedback. It should reduce users' cognitive load by enabling them to physically simulate processes and offload tasks like spatial perception and manipulation onto the body (Kirsh 2013). Distance and physical properties like size, shape, volume, weight, hardness, and texture can be automatically and subconsciously assessed with the body (Jeannerod 1997). Tangible interfaces should, therefore, enable users to subconsciously, kinesthetically judge and manipulate spatial distances, relationships, patterns, 3D forms, and volumes offloading these challenging cognitive tasks onto their bodies.

1.2   Tangible Geospatial Modeling

Tangible interfaces for geospatial modeling can transform the way we use GIS by affording intuitive, hands-on modes of embodied interaction, streamlining workflows for tasks like 3D modeling and analysis, and thus encouraging creative exploration. Embodied, tangible interaction should enhance users' spatial performance—their ability to sense, manipulate, and interact with multidimensional space—for challenging tasks like sculpting topography and guiding the flow of water by combining kinesthetic and computational affordances. Since tangible interfaces for geospatial modeling streamline workflows and enhance spatial performance, users can quickly develop new scenarios and quantitatively analyze the results in an analytical, yet creative process. There are already many tangible interfaces for geospatial modeling. These include shape changing interfaces (Table 1.1), augmented architectural interfaces (Table 1.2), augmented clay interfaces (Table 1.3), and augmented sandboxes (Table 1.4).

Shape changing interfaces (Rasmussen et al. 2012) or dynamic shape displays (Poupyrev et al. 2007) are a type of transformable tangible interface (Ishii et al. 2012). Typically these use motor-driven pistons to actuate an array of pins that physically change the shape of a tabletop surface based on computation. These tangible interfaces have three feedback loops—users can feel the physical model for passive, kinesthetic feedback, the model can be computationally transformed for active, kinesthetic feedback, and users can see computationally generated, graphical feedback.

Projection-augmented tangible interfaces rely on projection for representing digital data. Projected imagery has long been used to augment physical terrain models (Priestnall et al. 2012) (Fig. 1.2). Projection augmented tangible interfaces, however, are interactive. They couple physical and digital models through a cycle of 3D sensing or object recognition, computation, and projection. Augmented architectural interfaces like Urp (Underkoffler and Ishii 1999) and the Collaborative Design Platform (Schubert et al. 2011b) are a type of 'discrete tabletop tangible interface'

Table 1.1 Shape changing interfaces

System	Interaction	Studies	Publications
XenoVision Mark III Dynamic Sand Table	Sculpting		
Northrop Grumman Terrain Table	Sculpting		
Relief	Sculpting		Leithinger et al. (2009), Leithinger and Ishii (2010)
Recompose	Sculpting		Leithinger et al. (2011)
	Gesture		Blackshaw et al. (2011)
Tangible CityScape	Gesture		
inFORM	Sculpting		Follmer et al. (2013)
	Gesture		
	Object detection		

Table 1.2 Augmented architectural interfaces

System	Interaction	Studies	Publications
Urp	Object detection	Case studies*	Underkoffler and Ishii (1999), Ishii et al. (2002)*
Collaborative Design Platform	Object detection		Schubert et al. (2011b)
	Touch		Schubert et al. (2011a)
	Sketching		Schubert et al. (2012, 2014, 2015)
CityScope	Object detection		Hadhrawi and Larson (2016)

Note: symbols link type of study to relevant publications

Table 1.3 Augmented clay interfaces

System	Interaction	Studies	Publications
Illuminating Clay	Sculpting	Protocol analysis‡	Piper et al. (2002a,b), Fielding-Piper (2002), Shamonsky (2003)‡, Ishii et al. (2004), Ratti et al. (2004a)
Tangible Geospatial Modeling System	Sculpting	Case studies*	Mitasova et al. (2006), Tateosian et al. (2010)*

Note: symbols link type of study to relevant publications

Table 1.4 Augmented sandbox interfaces

System	Interaction	Studies	Publications
SandScape	Sculpting		Ishii et al. (2004), Ratti et al. (2004a)
PhoxelSpace	Sculpting		Ratti et al. (2004b)
Efecto Mariposa	Sculpting		Vivo (2011)
SandyStation	Sculpting		
Augmented Reality Sandbox	Sculpting	Survey[§]	Woods et al. (2016)[§]
Hakoniwa	Sculpting		Kikukawa et al. (2013)
	Gesture		
	Object detection		
	Sound		
Sedimachine	Physical simulation		Cantrell and Holzman (2014)
Rapid Landscape Prototyping Machine	Machining		Robinson (2014)
Tangible Landscape	Sculpting	Case studies*	Petrasova et al. (2014)
	Object detection	Quantitative experiments[†]	Petrasova et al. (2015)*
	Sketching		Harmon et al. (2016)[†], Harmon (2016)[†], Tabrizian et al. (2016, 2017)*, Harmon et al. (2018)[†], Millar et al. (2018)[†]
The Augmented REality Sandtable (ARES)	Sculpting	Quantitative experiments[†]	Amburn et al. (2015)
	Gesture		
Inner Garden	Sculpting		Schmidt-Daly et al. (2016)[†]
	Breathing		Roo et al. (2016)
	Emotion		

Note: symbols link type of study to relevant publications

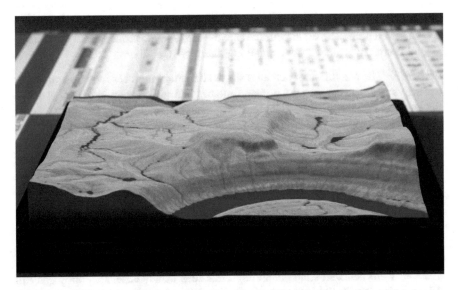

Fig. 1.2 A projection augmented model powered by Tangible Landscape with simulated water flow projected over 3D printed topography

(Ishii et al. 2012) with physical models of buildings that are augmented with projected analytics. Augmented clay interfaces like Illuminating Clay (Piper et al. 2002a) and augmented sandboxes like SandScape (Ishii et al. 2004) are types of 'deformable, continuous tangible interfaces' (Ishii et al. 2012) that users can sculpt. These tangible interfaces have two feedback loops—there is passive, kinesthetic feedback from grasping the physical model and active, graphical feedback from computation.

1.2.1 Shape Changing Interfaces

Shape changing interfaces—or dynamic shape displays—are computer controlled, interactive, physically responsive surfaces. As we interact with the physical surface it changes the digital model and, conversely, as we interact with the digital model the physical surface changes (Ishii 2008a; Poupyrev et al. 2007). Shape changing interfaces tend to be arrays of pistons and actuated pins that form kinetic, 2.5D surfaces (Petrie 2006) although there is experimental research into continuous, moving surfaces made of shape changing materials driven by heat, magnetic, or electrical stimuli (Coelho and Zigelbaum 2010).

Aegis Hyposurface The Aegis Hyposurface, an early example of a shape changing interface, is a generative art installation that uses pneumatic actuators to move a triangulated mesh surface according to an algorithm. It can be either preprogrammed or interactive, moving in response to sensed sound, light, or movement. As it was designed and built at an architectural scale the Aegis Hyposurface has a very coarse resolution for an actuated shape changing interface (Goulthorpe 2000).

FEELEX The resolution of actuated shape changing interfaces are constrained by the size and arrangement of the piston motors and piston rods or pins that move the surface. Project FEELEX, another early shape changing interface, used linear actuators to deform a rubber plate. The size of the motors—4 cm—meant that the resolution of the shape display was very coarse. Since the motors are larger than the pins, FEELEX 2 used a piston-crank mechanism to achieve a relatively high 8 mm resolution by clustering the pins while offsetting the motors below. A rubber sheet was stretched over the array of pins to create a 2.5D display for projection. When a user touched the surface they would depress the pins and the pressure of their touch would be recorded as a user interaction (Iwata et al. 2001).

Dynamic Sand Table and Terrain Table The XenoVision Mark III Dynamic Sand Table, developed in 2004, and the Northrop Grumman Terrain Table, developed in 2006, were actuated shape changing interfaces that represented topography in 2.5D. In the Terrain Table thousands of pins driven by a motor shaped a silicone surface held taut by suction from a vacuum below into a terrain. The Terrain Table recorded touches as user interactions such as panning and zooming. As users panned, zoomed, or loaded new geographic data, the actuated surface would automatically reshape within seconds (Petrie 2006).

Relief Relief is a relatively low-cost, scalable, 2.5D actuated shape display based on open source hardware and software. Given the complexity and thus the cost, maintenance, and unadaptability of earlier shape changing interfaces like FEELEX and the Northrop Grumman Terrain Table, Leithinger and Ishii (2010) aimed to design a simpler, faster system that was easier to build, adapt, scale, and maintain. In the first prototype of Relief an array of 120 actuated pins driven by electric slide potentiometers stretch a Lycra sheet into a shape display. Users can reshape the shape display by pressing or pulling on the actuated pins. The actuators are controlled with open source Arduino boards and a program written in the open source language Processing controls, senses, and tracks all of the pins and their relation to the digital model (Leithinger et al. 2009; Leithinger and Ishii 2010). The transparency and freedom of open source solutions should make it relatively easy to reconfigure and adapt this system.

Recompose While Relief was initially designed for a simple, highly intuitive interaction—direct physical manipulation (Leithinger and Ishii 2010)—its next iteration, Recompose, added gesture recognition (Leithinger et al. 2011; Blackshaw et al. 2011). While with Relief users can only directly sculpt the shape changing interface with their hands, with Recompose they can also use gestures to select, translate, rotate, and scale regions of the interface. The size and coarse resolution of the actuated interface mean that only small datasets or subsets of larger datasets can be modeled with useful fidelity. Furthermore, Leithinger et al. (2011) found that only a very limited range of touch interactions could be recognized at the same time and that it can be challenging to manipulate individual pins as they may be out of reach. They augment touch with gestures by adding a Kinect as a depth camera so that users can easily change the context and explore larger datasets. While gestures are less precise than direct physical manipulation, they greatly expand the scope of possible interactions (Blackshaw et al. 2011). Interactions via external devices such as a mouse may be less ambiguous than gestures, but Leithinger et al. argue that they draw users' focus away from the shape display. Therefore they choose to combine touch interactions with gestures rather than pointing devices so that the transition from sculpting to selection, translation, rotation, and scaling would be fluid and seamless given the physical directness of both modes of interaction (Leithinger et al. 2011).

Tangible CityScape Tangible CityScape, a system built upon Recompose, is an example of how this type of TUI can be applied to a specific domain—urban planning. It used a 2.5D shape changing interface to model and study urban massing. Building masses were modeled by clusters of pins and the model dynamically reshaped as users panned or zoomed with gestures (Tang et al. 2013).

inFORM With inFORM Follmer et al. (2013) developed a dynamically changing user interface capable of diverse, rich user interactions. Building on the Relief and Recompose systems, they developed a 2.5D actuated shape changing interface that supports object tracking, visualization via projection, and both direct and indirect physical manipulation. The surface of the interface is moved by a dense array of pins linked by connecting rods to a larger array of actuators below. The pins, pushing and pulling with variable pressure, offer nuanced haptic feedback to users. A Kinect depth sensor is used to track objects and users' hands. The actuated surface—a grid of pins—can be manipulated directly by pushing and pulling pins. Furthermore, users can interact with the system indirectly via object tracking. The surface can respond to interactions—both direct and indirect—and reshape itself. The actuated surface can also move objects placed on it, enabling indirect physical manipulations. Follmer et al. used this system to explore how shape-changing displays can dynamically model content and offer novel modes of interaction based on dynamically changing constraints and opportunities. As a responsive tangible user interface inFORM enables rich and varied mode of interaction such as responsive sculpting, moving passive objects, painting changes, and physically instantiated UI elements like buttons (Follmer et al. 2013).

Shape changing interfaces that can be shaped both by touch and computation enable physical interactions from human to computer and computer to human. With a shape changing interface the digital model is physically instantiated and this shape changing physical model is used for both input and output. When a user changes the surface of the shape display the computer reads the changes; when the computer changes the surface the user sees and feels the changes. Hypothetically this should radically reduce the level of abstraction in human-computer interaction. This novel mode of bidirectional, tangible interaction should be highly intuitive because it is so direct—human and computers communicate via the same medium. However, while dynamic shape displays may be highly intuitive, they have a relatively coarse resolution, are expensive, maintenance intensive, and hard to transport, all due to the actuators. The resolution of the display for example is constrained by the size of the actuators that make the display kinetic. The coarse resolution makes these displays' representations approximate and abstract, limiting their possible applications.

1.2.2 Augmented Architectural Interfaces

With augmented architectural interfaces users can place and move physical massing models of buildings, which are digitized using computer vision or 3D scanning. These interfaces enable users to intuitively model and visualize urban form and learn from computational feedback.

Urp Urp—a projection augmented interface for urban design—used tag-based objection detection to digitize physical models of buildings on a table. Spatial analyses and simulations such as proximity, wind, shadow, and viewsheds were computed and projected onto the tabletop in real-time so that users could rapidly test different spatial configurations of buildings (Underkoffler and Ishii 1999). As a case study Urp was used by a urban design class in the MIT School of Architecture and Planning. The researchers observed that it helped students to rapidly explore and test different configurations of space and effectively communicate their designs (Ishii et al. 2002).

Collaborative Design Platform The Collaborative Design Platform uses a depth camera to digitize and track physical models of buildings on a rear-projection light table. As users move polystyrene foam models of buildings, the models are 3D scanned updating a digital 3D model of a city. Analyzes like wind, light, shadow, accessibility, distance, and views are projected onto the table in realtime. Views are also rendered in 3D on a wall-mounted touch screen. Users can interact by placing and moving physical models of buildings, touching the screen, or sketching with a digitizing pen (Schubert et al. 2015).

CityScope MIT's City Science group developed CityScope, a tangible interface for urban modeling using Lego blocks. It is a participatory tool for urban planning, analysis, and prediction. With CityScope users build a Lego model of an urban neighborhood, which is 3D scanned to create a digital model of urban form. Then CityScope run simulations such as pedestrian and vehicular traffic, wind, and energy use and computes indices such as density, diversity, traffic, and proximity. Spatial data such as density, land use, and simulated traffic are projected onto the Lego model, while 3D facades, building temperature, and plots of indices can be rendered on a display. CityScope enables intuitive participatory urban modeling augmented with urban analytics. Users are able to rapidly explore different configurations of urban form, land use, and circulation and see the consequences (MIT Media Lab 2014; Hadhrawi and Larson 2016).

1.2.3 Augmented Clay Interfaces

Augmented clay interfaces couple a clay model of topography with a digital terrain model through a cycle of sculpting, 3D scanning, computation, and projection. As users sculpt the clay, the model is 3D scanned, the digital terrain model updates, and updated graphics are projected onto the clay model.

Illuminating Clay Illuminating Clay coupled a clay model and digital model of landscape through a cycle of laser scanning, spatial modeling, and projection. A clay model of a landscape was continuously scanned with a laser to generate a point cloud of x, y, and z coordinates which were then binned into a digital elevation model (DEM). The DEM or a derived topographic parameter such as slope, aspect, or cast shadow was then projected back onto the clay model so that users could see the impact of their changes in near real-time. Because Illuminating Clay used a laser scanner the scans were relatively fast—1.2 s each—and accurate to less than 1 mm, but the system was very expensive (Piper et al. 2002a,b). By enriching physical models of urban spaces and landscapes with spatial analyses such as elevation, aspect, slope, cast shadow, profile, curvature, viewsheds, solar irradiation, and water direction, Illuminating Clay enabled intuitive form-finding, streamlined analog and digital workflows, and enabled multiple users to simultaneously interact in a natural way (Ratti et al. 2004a). Illuminating Clay, however, had a very limited library of custom implemented spatial analyses. Since many of analyses were adapted from the open source GRASS GIS project (Piper et al. 2002a) there was a call for closer integration with GRASS GIS in order to draw on its extensive libraries for spatial computation (Piper et al. 2002b). The effort to couple a physical landscape model with GRASS GIS (Mitasova et al. 2006) led to the development of the Tangible Geospatial Modeling System (Tateosian et al. 2010).

Tangible Geospatial Modeling System The Tangible Geospatial Modeling System (TanGeoMS) coupled a physical model and GIS model of a landscape through a cycle of laser scanning, geospatial computation in GRASS GIS, and projection giving developers and users access to a sophisticated library for spatial modeling, simulation, visualization, and databasing in a highly intuitive environment. Like Illuminating Clay the system used a laser scanner to 3D scan a clay model. The scanned point cloud was then interpolated as a DEM, select geospatial analyses or simulations were computed from the DEM, and the results were projected back onto the clay model (Tateosian et al. 2010) (Fig. 1.3). It combined freeform hand modeling with geospatial modeling, simulation, and visualization so that users could easily explore how changes in topographic form affect landscape processes such as diffusive water flow and erosion and deposition (Mitasova et al. 2006).

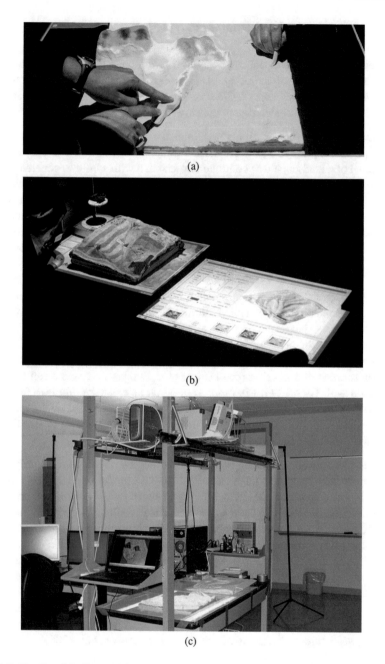

Fig. 1.3 The Tangible Geospatial Modeling System—an augmented clay interface powered by GRASS GIS: (**a**) multiple users interacting with a model, (**b**) projection augmented clay model coupled with a 3D rendering, (**c**) hardware setup with a laser scanner mounted above the model

1.2.4 Augmented Sandbox Interfaces

These tangible interfaces couple a sandbox with a digital terrain model through a cycle of sculpting, 3D scanning, computation, and projection. The sandbox is augmented with projected graphics such as simulated water flow. As users sculpt the sand, the sandbox is 3D scanned, the digital terrain model updates, and updated graphics are projected onto the sandbox.

SandScape SandScape used infrared depth sensing to digitize a 'sandbox' of 1 mm glass beads. An infrared camera captured the intensity of infrared light passing through the beads from below in real-time. A digital elevation model computed from the light intensity and derived analyses were projected back onto the sandbox for real-time feedback. SandScape was relatively low resolution due to the quality of the infrared sensing and the size of the glass beads (Ishii et al. 2004; Ratti et al. 2004a).

Phoxelspace Phoxel-Space adapted SandScape and Illuminating Clay for physically interacting with voxels, i.e. volumetric pixels or 3D rasters. The system coupled a malleable physical model—built of media like clay, plasticine, cubic blocks, or glass beads—with a 3D raster dataset using either a laser scanner or an IR camera. The researchers demonstrated how Phoxel-Space could be used to explore magnetic resonance imaging data, seismic velocity, and computational fluid dynamics (Ratti et al. 2004b).

Efecto Mariposa Efecto Mariposa was an interactive art installation using a Kinect sensor to scan a sand model of an island. The model of the island was augmented with a projection of a simulated ecosystem that changed as users sculpted the topography (Vivo 2011).

Augmented Reality Sandbox The Augmented Reality Sandbox developed by the UC Davis W.M. Keck Center for Active Visualization in the Earth Sciences couples a sandbox with a digital model of a landscape through a real-time cycle of 3D scanning with a Kinect sensor. As users sculpt the sand the Kinect sensor continually scans the sand surface generating a stream of depth maps. Scans are statistically filtered to remove hands and tools, reduce noise, and fill in areas with no data. The default filtering—30 frames—results in 1 s of lag. The digital elevation model, contours, and simulated water flow based on the shallow water equations are projected back onto the sand model of the landscape (Kreylos 2012). It was inspired by a Czech prototype called SandyStation developed in 2011 (Altman and Eckstein 2014). The code for this open source project, released under the GNU General Public License, and blueprints for building the system are available at https://arsandbox.ucdavis.edu. Two hundred eighty Augmented Reality Sandboxes have already been built around the world (Kreylos 2017).

Researchers at Eastern Carolina University built an Augmented Reality Sandbox and conducted a qualitative pilot study examining the effect of the technology on learning and engagement in geoscience education. Twelve students used the

Augmented Reality Sandbox to build terrain models from contours, model fluvial features and processes, and model coastal features and processes. The researchers solicited feedback with an exit survey. Students reported that the sandbox helped them learn about topography, fluvial and coastal processes, and process-form interactions more effectively. Based on the survey and their observations the researchers hypothesized that augmented sandboxes could enable embodied learning and encourage the development of scientific modeling skills (Woods et al. 2016).

Hakoniwa Hakoniwa—a projection-augmented sandbox for making generative music and art—was inspired by sandtray therapy. The system was designed to create a playful, embodied experience that could be therapeutic. As users built landscapes in the sandbox by sculpting sand and placing wooden blocks they created music and visual patterns in real-time through a cycle of depth and color sensing, image processing, audio generation, computer graphics, and projection (Kikukawa et al. 2013). The system evolved from PocoPoco—a tabletop tangible interface for making music (Kanai et al. 2011).

Rapid Landscape Prototyping Machine The Landscape Morphologies Lab at the University of Southern California developed the Rapid Landscape Prototyping Machine—a projection-augmented sandbox with robotic fabrication—to design and test strategies for dust control and mitigation for Lake Owens, California. The system used a 6-axis robotic arm to digitally fabricate algorithmically generated landscapes in a sandbox. The sand models were digitized with a laser scanner, the point cloud was triangulated as a terrain mesh, and spatial analyses such as viewsheds, aspect, and flooding were projected back onto the sandbox (Robinson 2014; Cantrell and Holzman 2016).

Augmented REality Sandtable The Augmented REality Sandtable (ARES) developed by the US Army Research Laboratory is designed for military training and simulation. This system uses a depth camera to continually 3D scan a sandbox and detect gestures. The digital elevation model and contour map derived from the scans and military units created using gestures or tablet input are projected onto the sand model. Units and buildings can be visualized in 3D using tablets or augmented reality glasses. The Augmented REality Sandtable can be linked with other military software to simulate scenarios (Amburn et al. 2015). A user study comparing users' performance with paper maps, Google Earth, and the Augmented REality Sandtable found that the sandtable was the most effective technology. Participants—especially participants who were veteran video gamers—tended to perform better with the sandtable in landmark identification, distance estimation, and situational judgment tests (Schmidt-Daly et al. 2016).

Inner Garden Inner Garden—a projection-augmented sandbox for contemplation and self-reflection—couples a sand model of topography with simple digital environment with water, plants, clouds, and daylight through a cycle of 3D scanning with a Kinect, biometric sensing, computation, and projection. While a user sculpts topography in the sand creating a digital elevation model, their physiological and emotional state are monitored with an electroencephalogram (EEG) and breathing sensor. Their breath controls the simulated sea level and daylight, their level of frustration controls the simulated cloud cover, and their meditativeness controls simulated plant life (Roo et al. 2016).

1.3   Tangible Landscape

Tangible Landscape is a projection-augmented sandbox powered by a GIS for real-time geospatial analysis and simulation (Petrasova et al. 2015). It was designed to intuitively 3D sketch landscapes—to rapidly exploring ideas or test hypotheses with real-time computational feedback (Fig. 1.4). It evolved from Illuminating Clay (Piper et al. 2002a) and the Tangible Geospatial Modeling System (Tateosian et al. 2010). While the Tangible Geospatial Modeling System used an expensive laser scanner for 3D sensing (Tateosian et al. 2010), Tangible Landscape—inspired by the open source Augmented Reality Sandbox (Kreylos 2012)—uses a low-cost 3D sensor for real-time depth and color sensing. The 1st generation of Tangible Landscape (Petrasova et al. 2014) used the 1st generation Kinect with structured light sensing (Smisek et al. 2011), while the 2nd (Petrasova et al. 2015) and 3rd generations of Tangible Landscape used the 2nd generation Kinect with time-of-flight sensing (Bamji et al. 2015). Tangible Landscape is tightly integrated with GIS, using a GRASS GIS plugin to automatically scan, process, georeference, import, and analyze the model. Because it is so tightly integrated with GRASS GIS users can also use the GUI, the command line, and scripting as advanced controls for tasks not suited to a TUI (Petrasova et al. 2014). As a projection-augmented sandbox with 3D sensing and color and object recognition a wide range of media can be used such as polymer-enriched sand, clay, 3D prints, CNC-machined models, architectural models, colored felt, and wooden markers.

Conceptually, Tangible Landscape couples a physical model with a digital model in a real-time feedback cycle of 3D scanning, geospatial modeling and simulation, and projection and 3D rendering. For example, by sculpting the terrain of the physical model, we can see how the changes affect processes like the flow of water, flooding, erosion, and solar irradiation. Thus we can easily and rapidly test ideas while being guided by scientific feedback, exploring a much a larger solution space and make more creative and informed decisions. And since many users can interact with the physical model at once, Tangible Landscape encourages collaboration and interdisciplinary exchange (Fig. 1.5). Tangible Landscape combines real-time interaction with extensive scientific tools for modeling, analysis, and visualization at the precision needed for real-world design and planning applications. Tangible Landscape can be used for applications such as grading landforms (Chap. 3), analyzing topography (Chap. 6), modeling water flow and soil erosion (Chap. 7), analyzing viewsheds (Chap. 9), planning trail networks (Chap. 10), analyzing solar radiation dynamics (Chap. 11), simulating and managing fire (Chap. 12), modeling inundation and flooding (Chap. 13), and landscape design (Chap. 14).

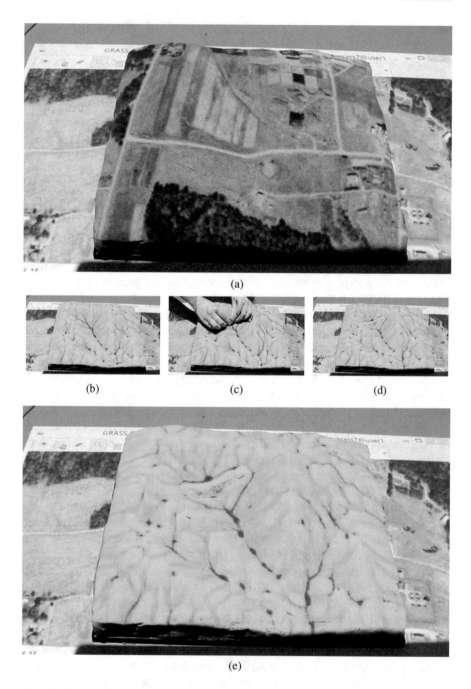

Fig. 1.4 3D sketching a check dam with Tangible Landscape: (**a**) orthoimagery projected over the model, (**b**) water flow, (**c**) sculpting a check dam, (**d**) updated water flow, (**e**) a dammed valley with water flow simulated with *r.sim.water*

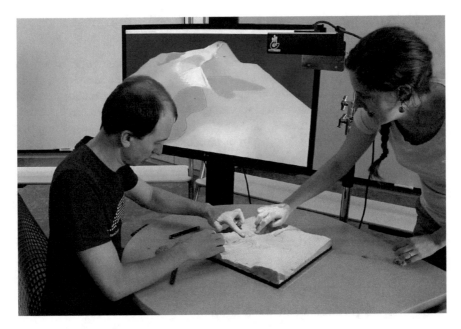

Fig. 1.5 Collaboratively sculpting a lake with Tangible Landscape

Tangible Landscape is inexpensive, portable, adaptable, and easy to implement. As a high resolution sandbox-style tangible interface it can be used for both rapid ideation and precise modeling. However, unlike a shape changing interface the physical model is passive and cannot be changed computationally in real-time—trading real-time response for high resolution, high precision, simplicity of setup, and low cost. Rapid prototyping can be used to precisely shape or reshape the physical model allowing users to work with real world landscapes and easily reset the conditions.

Tangible Landscape has been assessed in a number user studies. Studies by Harmon et al. found that users were able to sculpt more accurate topographic models with more distinct landforms using Tangible Landscape than they were using digital 3D modeling or analog modeling by hand. The study also found that users worked in a rapid, iterative process learning from real-time geospatial analytics (Harmon et al. 2016; Harmon 2016; Harmon et al. 2018). Millar et al. studied the effectiveness of Tangible Landscape as a tool for teaching about grading (i.e. earthwork), geomorphology, and hydrology. This study found that tangible teaching methods were highly engaging, enabled a natural learning process, and helped students build task-specific knowledge about topics such as "cut and fill" (Millar et al. 2018).

1.3.1   Developing Tangible Landscape

Tangible Landscape is a free, open source project with source code hosted on GitHub.[1] The system is constantly evolving in response to new developments in its hardware and software components and the needs of its expanding range of applications. Enhanced point cloud processing, optimization of the core functions, and migration to more efficient hardware and operating systems will improve real-time interaction and allow us to incorporate more sophisticated dynamic modeling. User interaction and experience can be further improved with dashboards and steered simulations. Computationally controlled shape generation with in-situ digital fabrication will combine the advantages of dynamic and continuous shape displays. By providing Tangible Landscape as a free, open source project, along with this book, we aim to build a community of developers and users of this technology for the benefit of education, research and collaborative decision making in communities worldwide.

1.4   The Organization of This Book

This book describes tangible geospatial modeling using open source GIS and its applications. For each application we explain the underlying theory and algorithms, provide the workflows used, and describe case studies. The workflows include GRASS GIS and Rhinoceros command calls as well as GRASS GIS and Blender Python code snippets.

Chapter 2 describes how Tangible Landscape works and how it is set up. This chapter details the 3D scanning technology, the hardware and software used, and the physical setup of the system. It provides information about where to download GRASS GIS and the Tangible Landscape plugin and explains how to read and use the workflows. Chapter 3 explains how to build physical models for use with Tangible Landscape. This chapter covers handmade modeling techniques, digital fabrication techniques, and techniques for molding and casting models. Chapter 4 introduces and explains the different modes of tangible interaction. Chapter 5 covers 3D visualization, explaining how to 3D model and 3D render scenes in real-time with Tangible Landscape. Chapter 6 covers the basic processes of scanning, interpolating, and analyzing a terrain model. The following chapters describe a range of scientific applications for Tangible Landscape. Each chapter first covers the relevant theory and algorithms and then demonstrates the application in a case study. Chapter 7 covers surface water flow modeling. Chapter 8 covers soil erosion modeling. Chapter 9 covers viewshed analysis. Chapter 10 covers trail planning. Chapter 11 covers solar irradiation dynamics. Chapter 12 covers simulating the

---

[1]https://github.com/tangible-landscape/grass-tangible-landscape.

spread of wildfire. Chapter 13 covers coastal inundation and flooding modeling. Chapter 14 covers landscape architecture and design. Appendix includes case studies of additional applications of Tangible Landscape, a summary of the data used in this book with download links, a list of useful online data sources, and a guide to getting started with GRASS GIS.

References

Altman, P., & Eckstein, R. (2014). SandyStation [online]. Accessed 12.08.2015. http://en.sandystation.cz/.

Amburn, C. R., Vey, N. L., Boyce, M. W., & Mize, J. R. (2015). The Augmented REality Sandtable (ARES). Technical Report October. US Army Research Laboratory.

Anderson, M. L. (2008). Evolution, embodiment and the nature of the mind. In B. Hardy-Vallee & N. Payette (Eds.), *Beyond the brain*, chapter 1. Newcastle, UK: Cambridge Scholars Publishing.

Bamji, C. S., O'Connor, P., Elkhatib, T., Mehta, S., Thompson, B., Prather, L. A., Snow, D., Akkaya, O. C., Daniel, A., Payne, A. D., Perry, T., Fenton, M., & Chan, V. H. (2015). A 0.13 μm CMOS system-on-chip for a 512 x 424 time-of-flight image sensor with multi-frequency photo-demodulation up to 130 MHz and 2 GS/s ADC. *IEEE Journal of Solid-State Circuits, 50*(1), 303–319.

Blackshaw, M., DeVincenzi, A., Lakatos, D., Leithinger, D., & Ishii, H. (2011). Recompose: Direct and gestural interaction with an actuated surface. In *Proceedings of the 2011 Annual Conference Extended Abstracts on Human Factors in Computing Systems - CHI EA '11* (p. 1237). Vancouver: ACM Press.

Cantrell, B., & Holzman, J. (2014). Synthetic ecologies: Protocols, simulation, and manipulation for indeterminate landscapes. In *ACADIA 14: Proceedings of the 34th Annual Conference of the Association for Computer Aided Design in Architecture*, Los Angeles (pp. 709–718).

Cantrell, B., & Holzman, J. (2016). Rapid landscape prototyping machine. In *Responsive Landscapes*, chapter 6.6 (pp. 159–166). New York: Routledge.

Coelho, M., & Zigelbaum, J. (2010). Shape-changing interfaces. *Personal and Ubiquitous Computing, 15*(2), 161–173.

Dourish, P. (2001). *Where the action is: The foundations of embodied interaction*. Cambridge, MA: MIT.

Fielding-Piper, B. T. (2002). The illuminated design environment: A 3-D tangible interface for landscape analysis. Master's thesis, Massachusetts Institute of Technology.

Fitzmaurice, G. W., Ishii, H., & Buxton, W. (1995). Bricks: Laying the foundations for graspable user interfaces. In *Proceedings of the SIGCHI Conference on Human Factors in Computing Systems* (pp. 442–449).

Follmer, S., Leithinger, D., Olwal, A., Hogge, A., & Ishii, H. (2013). inFORM: Dynamic physical affordances and constraints through shape and object actuation. In *UIST '13 Proceedings of the 26th Annual ACM Symposium on User Interface Software and Technology* (pp. 417–426). St. Andrews, UK: ACM Press.

Goulthorpe, M. (2000). Aegis Hyposurface [online]. Accessed 10.04.2015. http://hyposurface.org/.

Hadhrawi, M., & Larson, K. (2016). Illuminating legos with digital information to create urban data observatory and intervention simulator. In *Proceedings of the 2016 ACM Conference Companion Publication on Designing Interactive Systems*, DIS '16 Companion (pp. 105–108). New York, NY, USA: ACM.

Harmon, B. A. (2016). Embodied spatial thinking in tangible computing. In *Proceedings of the TEI '16: Tenth International Conference on Tangible, Embedded, and Embodied Interaction*, TEI '16 (pp. 693–696). New York, NY, USA: ACM.

Harmon, B. A., Petrasova, A., Petras, V., Mitasova, H., & Meentemeyer, R. (2018). Tangible topographic modeling for landscape architects. *International Journal of Architectural Computing, 16*, 4–21.

Harmon, B. A., Petrasova, A., Petras, V., Mitasova, H., & Meentemeyer, R. K. (2016). Tangible landscape: Cognitively grasping the flow of water. *International Archives of the Photogrammetry, Remote Sensing and Spatial Information Sciences, XLI-B2*, 647–653.

Ishii, H. (2008a). Tangible bits: Beyond pixels. In *Proceedings of the 2nd International Conference on Tangible and Embedded Interaction - TEI '08* (pp. xv–xxv). Bonn, Germany. ACM Press.

Ishii, H. (2008b). The tangible user interface and its evolution. *Communications of the ACM, 51*(6):32–36.

Ishii, H., Lakatos, D., Bonanni, L., & Labrune, J.-B. (2012). Radical atoms: Beyond tangible bits, toward transformable materials. *Interactions, 19*(1), 38–51.

Ishii, H., Ratti, C., Piper, B., Wang, Y., Biderman, A., & Ben-Joseph, E. (2004). Bringing clay and sand into digital design—continuous tangible user interfaces. *BT Technology Journal, 22*(4), 287–299.

Ishii, H., & Ullmer, B. (1997). Tangible bits: Towards seamless interfaces between people, bits and atoms. In *Proceedings of the SIGCHI Conference on Human Factors in Computing Systems - CHI '97* (pp. 234–241). New York, USA: ACM Press.

Ishii, H., Underkoffler, J., Chak, D., Piper, B., Ben-Joseph, E., Yeung, L., & Kanji, Z. (2002). Augmented urban planning workbench: Overlaying drawings, physical models and digital simulation. In *ISMAR '02 Proceedings of the 1st International Symposium on Mixed and Augmented Reality* (pp. 203–211). Washington, DC: IEEE Computer Society.

Iwata, H., Yano, H., Nakaizumi, F., & Kawamura, R. (2001). Project FEELEX: Adding haptic surface to graphics. In *Proceedings of SIGGRAPH 2001* (pp. 469–475).

Jeannerod, M. (1997). *The cognitive neuroscience of action.* Cambridge, MA: Blackwell.

Just, M. A., & Carpenter, P. A. (1985). Cognitive coordinate systems: Accounts of mental rotation and individual differences in spatial ability. *Psychological Review, 92*(2), 137–172.

Kanai, T., Kikukawa, Y., Suzuki, T., Baba, T., & Kushiyama, K. (2011). Pocopoco: A tangible device that allows users to play dynamic tactile interaction. In *ACM SIGGRAPH 2011 Emerging Technologies*, SIGGRAPH '11 (pp. 12:1–12:1), New York, NY, USA: ACM.

Kikukawa, Y., Kato, M., Baba, T., & Kushiyama, K. (2013). Hakoniwa: A sonification art installation consists of sand and woodblocks. In *Proceedings of the 19th International Conference on Auditory Display (ICAD 2013)* (pp. 283–286). Lodz, Poland: International Community for Auditory Display.

Kirsh, D. (2013). Embodied cognition and the magical future of interaction design. *ACM Transactions on Computer-Human Interaction, 20*(1), 3:1–3:30.

Kreylos, O. (2012). Augmented Reality Sandbox [online]. Accessed 20.01.2017. http://idav.ucdavis.edu/~okreylos/ResDev/SARndbox/index.html

Kreylos, O. (2017). Augmented Reality Sandbox [online]. https://arsandbox.ucdavis.edu/

Leithinger, D., & Ishii, H. (2010). Relief: A scalable actuated shape display. In *Proceedings of the Fourth International Conference on Tangible, Embedded, and Embodied Interaction - TEI '10* (p. 221). Cambridge, MA: ACM Press.

Leithinger, D., Kumpf, A., & Ishii, H. (2009). Relief [online]. http://tangible.media.mit.edu/project/relief/

Leithinger, D., Lakatos, D., Devincenzi, A., Blackshaw, M., & Ishii, H. (2011). Direct and gestural interaction with relief: A2. 5D shape display. In *Proceedings of the 24th Annual ACM Symposium on User Interface Software and Technology* (pp. 541–548).

Millar, G. C., Tabrizian, P., Petrasova, A., Petras, V., Harmon, B., & Meentemeyer, R. K. (2018). Tangible landscape: A hands-on method for teaching terrain analysis. In *CHI '18 Proceedings of the 2018 CHI Conference on Human Factors in Computing Systems*, Montreal, Canada.

MIT Media Lab (2014). CityScope. http://cp.media.mit.edu/cityscope/

Mitasova, H., Mitas, L., Ratti, C., Ishii, H., Alonso, J., & Harmon, R. S. (2006). Real-time landscape model interaction using a tangible geospatial modeling environment. *IEEE Computer Graphics and Applications, 26*(4), 55–63.

Petrasova, A., Harmon, B., Petras, V., & Mitasova, H. (2015). *Tangible modeling with open source GIS*. Berlin: Springer.

Petrasova, A., Harmon, B. A., Petras, V., & Mitasova, H. (2014). GIS-based environmental modeling with tangible interaction and dynamic visualization. In D. Ames & N. Quinn (Eds.), *Proceedings of the 7th International Congress on Environmental Modelling and Software*, San Diego, California, USA. International Environmental Modelling and Software Society.

Petrie, G. (2006). TouchTable & TerrainTable - showstoppers at the ESRI user conferences. *Geoinformatics, 9*(2), 40–41.

Piper, B., Ratti, C., & Ishii, H. (2002a). Illuminating clay: A 3-D tangible interface for landscape analysis. In *Proceedings of the SIGCHI Conference on Human Factors in Computing Systems - CHI '02* (p. 355), Minneapolis: ACM Press.

Piper, B., Ratti, C., & Ishii, H. (2002b). Illuminating clay: A tangible interface with potential GRASS applications. In *Proceedings of the Open Source GIS - GRASS Users Conference 2002*, Trento, Italy.

Poupyrev, I., Nashida, T., & Okabe, M. (2007). Actuation and tangible user interfaces: The Vaucanson duck, robots, and shape displays. In *Proceedings of TEI 2007* (pp. 205–212).

Poynor, R. (1995). The hand that rocks the cradle. *ID Magazine, 42*, 60–65.

Priestnall, G., Gardiner, J., Way, P., Durrant, J., & Goulding, J. (2012). Projection Augmented Relief Models (PARM): Tangible displays for geographic information. In *Proceedings of Electronic Visualisation and the Arts*, London (pp. 1–8).

Rasmussen, M. K., Pedersen, E. W., Petersen, M. G., & Hornbæk, K. (2012). Shape-changing interfaces: A review of the design space and open research questions. In *Proceedings of the SIGCHI Conference on Human Factors in Computing Systems*, CHI '12 (pp. 735–744). New York, NY, USA. ACM.

Ratti, C., Wang, Y., Ishii, H., Piper, B., Frenchman, D., Wilson, J. P., Fotheringham, A. S., & Hunter, G. J. (2004a). Tangible User Interfaces (TUIs): A novel paradigm for GIS. *Transactions in GIS, 8*(4), 407–421.

Ratti, C., Wang, Y., Piper, B., Ishii, H., & Biderman, A. (2004b). Phoxel-space: An interface for exploring volumetric data with physical voxels. In *Proceedings of the 5th Conference on Designing Interactive Systems: Processes, Practices, Methods, and Techniques*, DIS '04, pages 289–296, New York, NY, USA. ACM.

Robinson, A. (2014). Calibrating agencies in a territory of instrumentality: Rapid landscape prototyping for the owens lake dust control project. In *ACADIA 14: Projects of the 34th Annual Conference of the Association for Computer Aided Design in Architecture (ACADIA)* (pp. 143–146).

Roo, J. S., Gervais, R., & Hachet, M. (2016). Inner garden: An augmented sandbox designed for self-reflection. In *Proceedings of the TEI '16: Tenth International Conference on Tangible, Embedded, and Embodied Interaction*, TEI '16 (pp. 570–576). New York, NY, USA: ACM.

Schmidt-Daly, T. N., Riley, J. M., Amburn, C. R., Hale, K. S., & Yacht, P. D. (2016). Video game play and effect on spatial knowledge tasks using an augmented sand table. *Proceedings of the Human Factors and Ergonomics Society Annual Meeting, 60*(1), 1429–1433.

Schubert, G., Artinger, E., Petzold, F., & Klinker, G. (2011a). Bridging the gap: A (collaborative) design platform for early design stages. In *Education and Research in Computer Aided Architectural Design in Europe*, Ljubljana (Vol. 29, pp. 187–193).

Schubert, G., Artinger, E., Petzold, F., & Klinker, G. (2011b). Tangible tools for architectural design: Seamless integration into the architectural workflow. In *Proceedings of Association for Computer Aided Design in Architecture*, Banff, Canada (pp. 1–12).

Schubert, G., Artinger, E., Yanev, V., Petzold, F., & Klinker, G. (2012). 3D Virtuality sketching: Interactive 3D-sketching based on real models in a virtual scene. *Proceedings of the 32nd Annual Conference of the Association for Computer Aided Design in Architecture, 32*, 409–418.

Schubert, G., Schattel, D., Marcus Tönnis, G. K., & Petzold, F. (2015). Tangible mixed reality on-site: interactive augmented visualisations from architectural working models in urban design.

In *Computer-aided architectural design futures. the next city - new technologies and the future of the built environment* (Vol. 527, pp. 55–74). Berlin, Heidelberg: Springer.

Schubert, G., Tönnis, M., Yanev, V., Klinker, G., & Petzold, F. (2014). Dynamic 3d-sketching. *Proceedings of the 19th International Conference on Computer-Aided Architectural Design Research in Asia, 19*, 107–116.

Shamonsky, D. J. (2003). Tactile, spatial interfaces for computer-aided design: superimposing physical media and computation. PhD thesis, Massachusetts Institute of Technology.

Shepard, R. N., & Metzler, J. (1971). Mental rotation of three-dimensional objects. *Science, 171*(3972), 701–703.

Smisek, J., Jancosek, M., & Pajdla, T. (2011). 3D with Kinect. In *Proceedings of the IEEE International Conference on Computer Vision* (pp. 1154–1160). Barcelona: IEEE.

Tabrizian, P., Harmon, B. A., Petrasova, A., Mitasova, H., & Meentemeyer, R. K. (2017). Tangible immersion for ecological design. In *ACADIA 17: Proceedings of the 37th Annual Conference of the Association for Computer Aided Design in Architecture*, Cambridge, MA (pp. 600–609)

Tabrizian, P., Petrasova, A., Harmon, B., Petras, V., Mitasova, H., & Meentemeyer, R. (2016). Immersive tangible geospatial modeling. In *Proceedings of the 24th ACM SIGSPATIAL International Conference on Advances in Geographic Information Systems - GIS '16* (pp. 1–4).

Tang, S. K., Sekikawa, Y., Leithinger, D., Follmer, S., & Ishii, H. (2013). Tangible CityScape [online]. Accessed 27.03.2014. http://tangible.media.mit.edu/project/tangible-cityscape/

Tateosian, L., Mitasova, H., Harmon, B. A., Fogleman, B., Weaver, K., & Harmon, R. S. (2010). TanGeoMS: Tangible geospatial modeling system. *IEEE transactions on visualization and computer graphics, 16*(6), 1605–1612.

Underkoffler, J., & Ishii, H. (1999). Urp: A luminous-tangible workbench for urban planning and design. In *CHI '99 Proceedings of the SIGCHI Conference on Human Factors in Computing Systems* (pp. 386–393). New York, New York, USA. ACM Press.

Uttal, D. H., Miller, D. I., Newcombe, N. S. (2013). Exploring and enhancing spatial thinking: links to achievement in science, technology, engineering, and mathematics? *Current Directions in Psychological Science, 22*(5), 367–373.

Vivo, P. G. (2011). Efecto Mariposa. http://patriciogonzalezvivo.com/2011/efectomariposa/

Woods, T. L., Reed, S., Hsi, S., Woods, J. A., & Woods, M. R. (2016). Pilot study using the augmented reality sandbox to teach topographic maps and surficial processes in introductory geology labs. *Journal of Geoscience Education, 64*(3), 199–214.

Chapter 2
System Configuration

The setup of the Tangible Landscape system consists of four primary components: (a) a physical model that can be modified by a user, (b) a 3D scanner, (c) a projector, and (d) a computer installed with GRASS GIS for geospatial modeling and additional software that connects all the components together. The physical model, placed on a table, is scanned by the 3D scanner mounted above. The scan is then imported into GRASS GIS, where a DEM is created. The DEM is then used to compute selected geospatial analyses. The resulting image or animation is projected directly onto the modified physical model so that the results are put into the context of the modifications to the model.

2.1 Hardware

Tangible Landscape can be built with affordable, commonly available hardware: a 3D scanner, a projector and a computer. We describe some of the current hardware options while noting that the technology develops rapidly and alternative, more effective solutions may emerge.

2.1.1 3D Scanner

In the Tangible Landscape workflow the 3D scanner captures the physical model as it is modified by users. Therefore it is important that the device scans as accurately as the technology allows. The Tangible Landscape system was developed using the Kinect for Xbox One (formerly Kinect for Windows v2), which is one of the most affordable 3D scanners on market, providing real-time, high-resolution depth and

© The Author(s) 2018
A. Petrasova et al., *Tangible Modeling with Open Source GIS*,
https://doi.org/10.1007/978-3-319-89303-7_2

color information. Due to the growing demand this technology is developing rapidly and new, improved 3D scanners may become available soon.

The Technology Behind 3D Scanning We describe the basic principles behind current 3D depth sensing technologies in order to explain both the potential and the limitations of current scanning devices. More detailed information can be found for example in Mutto et al. (2012).

Several currently available depth sensors such as Apple Primesense Carmine, Asus Xtion PRO LIVE,[1] and Kinect for Windows v1, are based on a Primesense proprietary light-coding technique. This technique uses triangulation to map 3D space in manner similar to how the human visual system senses depth from two slightly different images. Rather than using two cameras it triangulates between a near-infrared camera and a near-infrared laser source. Corresponding objects need to be identified in order to triangulate between the images. A light coding (or structured light) technique is used to identify the objects. The laser produces a pseudo-random dot pattern which is then used to find the matching dot pattern in the projected pattern. In this way the final depth image can be constructed.

Kinect for Xbox One uses *Time-of-Flight* (ToF), a technique widely used in lidar technology. It has a sensor that indirectly measures the time it takes for pulses of near-infrared laser light to travel from a laser source, to a target surface, and then back to the image sensor. Time-of-Flight sensors are generally considered to be more precise, but also more expensive.

Since the scanning and 3D modeling algorithms and their implementations for most sensors are proprietary, the specific behavior and precision of particular sensors is subject to many experimental studies. Generally, sensors using near-infrared light are sensitive to lighting conditions, so outdoor usage is typically not recommended. The depth resolution decreases with the distance from the sensor and also at close range (i.e. between a couple of millimeters to tens of centimeters). The range and field of view of the sensors can vary; Tangible Landscape requires short range sensors with a minimum distance of 0.5 m to keep the highest possible resolution. When scanning with one sensor the size of the physical model is limited by the required accuracy because the sensor must be far enough away to capture the entire model in the sensor's field of view, which for the Kinect for Xbox One is 60° vertical by 70° horizontal.

Detailed information about accuracy and precision of Kinect for Windows v1 and Kinect for Xbox One can be found in Wasenmüller and Stricker (2017), Andersen et al. (2012), Lachat et al. (2015) and a comparison to Asus Xtion can be found in Gonzalez-Jorge et al. (2013). The main limiting factors of Kinect for Xbox One accuracy include the correlation of depth accuracy and temperature of the scanner, influence of scene color on depth estimation, flying pixels (erroneous pixels, a well-known artifact of ToF cameras) along depth discontinuities, and high depth deviation in image corners. Knowing these limitations, we can to certain extent

---

[1] https://www.asus.com/us/3D-Sensor/Xtion_PRO_LIVE/.

compensate for them by pre-heating the scanner before measuring, avoiding high contrast scenes, and using statistical filtering methods to avoid flying pixels (see Sect. 2.2.3 for more details).

2.1.2 Projector

The projector projects the background geospatial data and results of analyses onto the 3D physical model. Therefore it is important to select a projector with sufficient resolution and properties that minimize distortion and generate a bright image.

Resolution and brightness are two important criteria to be considered. We recommend higher resolution projectors offering at least WXGA (1280×800). The brightness depends upon where Tangible Landscape is used and whether the room's ambient light can be reduced for the sake of brighter projected colors. We recommend brighter projectors (at least 3000 lumens) since we project on a variety of materials which are not always white and reflective.

There are two important types of projectors—standard and short-throw projectors. They differ in throw ratio values, which are defined as the distance from the projector's lens to the screen, divided by the width of the projected image (short throw and ultra short throw projectors have ratios 0.3–0.7, while the standard projectors have throw ratio values around 2). At the same distance the projector with a lower throw ratio can display a larger image. In other words a projector with a lower throw ratio (short-throw) can project an image of the same size as the higher throw ratio projector, but from a shorter distance. Tangible Landscape can be set up using both types of projectors; each has advantages and disadvantages.

The placement and configuration of the projector is important because it affects the coverage, distortion, and visibility of the projected data. For example in some setups the 3D scanner may be caught in the projection and would thus cast a shadow over the model. Therefore it can be practical to use a short-throw projector because it can be placed to the side of the physical model at a height similar to the 3D scanner (Figs. 2.1b and 2.3). Since the projection is cast from the side the projection beam does not cross the 3D scanning device and no shadow is cast. However, with a short-throw projector there is a certain level of visible distortion when projecting on a physical model that has substantial relief. The distortion occurs because the light rays reach the model at a very acute angle. The horizontal position at which the projected light intersects the model is shifted from the position at which it would intersect with a flat surface. Larger differences in height increase the distortion. Theoretically we can remove the distortion by either warping the projected data itself or using the projector to automatically warp the projected image. The first solution would require an undistorted dataset for geospatial modeling and a warped copy of that data for projecting. That is impractical especially when working with many different raster and vector layers. The other solution requires the projector itself to warp the image; while this technology exists, it is only offered by a few projector manufacturers and such projectors are typically more expensive.

Moreover, as the landscape is modified, the warping pattern should change as well. Currently it is not possible to find this feature in the off-the-shelf projectors.

With standard projectors the distortion is usually negligible since the incidence angle of the rays is relatively small. A standard projector needs to be placed much higher above the model than a short-throw projector does due to the difference in throw ratios (Fig. 2.1a). This creates several challenges. It is hard to mount and manipulate the projector when is it so high above the model. Furthermore it is challenging to align projector and the 3D scanner without casting a shadow. Since the 3D scanner has a limited field of view it must be placed close to the horizontal center of the model. When the projector is mounted above the scanner, the scanner is caught in the projector's beam, casting a shadow over part of the model. For small to mid sized models, depending on the particular setup and the height of the scanner and projector, this may not be a problem.

The shadow of the scanner can also be avoided with specialized short throw projectors that allow greater installation flexibility through lens shifting. These projectors (for example Canon WUX400ST) are capable of projecting from the horizontal center of the projected area at the same height as the scanner (Fig. 2.1c). A device which combines the scanner and projector would make this setup easy; an appropriate device, however, was not available at the time of writing.

To minimize the distortion when using projectors with lower throw ratios, we can project from the center by tilting the projector and correcting the resulting keystone distortion (Fig. 2.1d). With this setup we recommend projectors that have throw ratios around 1.0 (for example Epson PowerLite 1700 Series) as a lower ratio can result in additional distortion. When testing the projector setup it is useful to project the rectangular grid on a flat surface in order to quickly check if there is any distortion.

2.1.3 Computer Requirements

System requirements depend largely upon the sensor and its associated library or software development kit (SDK). Certain sensors, such as Kinect for Xbox, are designed to work on specific operating systems with the producer's SDK, however open source drivers, namely *libfreenect2* (Xiang et al. 2016), allow users to run Kinect on other platforms. The preferred platform of Tangible Landscape is GNU/Linux distribution Ubuntu, see notes in Sect. 2.2.5. The computer should be configured for 3D scanning and geospatial modeling, both of which are performance- and memory-intensive processes. The hardware requirements are very similar to the requirements for gaming computers: a multi-core processor, at least 4 GB of system memory and a good graphics card are necessary to achieve real-time interaction with the model. The specific parameters required for the scanner device should be reviewed on the manufacturer's website.

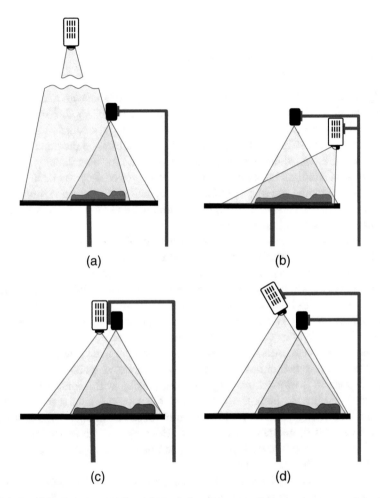

Fig. 2.1 Possible placements of the projector and scanner according to projector type: (**a**) standard projector mounted in the ceiling, (**b**) standard short throw projector, (**c**) short throw projector capable of projecting above the center, (**d**) short throw projector tilted to project above the center with keystone correction

2.1.4   *Physical Setup*

The setup for Tangible Landscape system is quite flexible. However, there are some crucial components and some specifics that make a setup more usable. When building Tangible Landscape in the laboratory or when bringing it to the community the following items are necessary:

- a table for a model
- a laptop or desktop computer with a table

- a scanner
- a projector
- 1–2 stands for a projector and a scanner
- 3–4 power plugs (and/or extension cable)

Ideally the table should be either a 90 cm × 90 cm square table with rounded corners, a rounded table 100 cm in diameter, or a teardrop table of similar size. To freely interact with the model a person should be able to almost reach the other side of the table; this is not possible with larger tables. Smaller tables, on the other hand, have less space for tools, additional sand, and application windows. Ideally application windows with additional information should be projected onto the table beside the model. A large model can be placed on top of a smaller table or stand, provided that this base is stable. In this case if any application windows are needed, they have to be projected directly onto the model.

A round table is quite advantageous because people can stand at equal distances from the model and can easily move around the table. Unfortunately, one side of the table is typically occupied by the metal stands for projector and scanner since ceiling mounts are rarely possible and render the system immobile. The table should be stable enough to hold sand and models and sturdy enough to withstand their manipulation. We recommend putting the computer on a separate table so that the modeling table is not cluttered. The more space there is around the modeling table, the more access users have to the model. The computer, however, needs to be situated so that its operator has easy access to the model as well.

An alternative is to use a rectangular table (ideally with rounded corners) that is 140 cm × 90 cm for both the model and the laptop. A narrow table that is 150 cm × 75 cm may work as well. This setup makes the model less accessible as at least two sides of the model are blocked. However, the whole model should be accessible from any of the remaining sides, so this setup does not necessarily limit interaction, but rather the number of users.

Sometimes it is advantageous to have a large screen showing additional data or 3D views to enhance users' understanding of the processes. However, such screen should not limit access to the physical model. Tangible Landscape could be also extended with additional devices such as 3D displays, head-mounted displays, and hand-held devices.

There should be at least 80 cm of space on each side of the modeling table so that there is room for walking, standing, and placing the stands. The whole area required is about 2.5 m × 2.5 m but in practice the area is often larger depending on the size of the tables and the surroundings. Generally, it is better to have a larger open space so that people can spread around the different sides of Tangible Landscape setup.

The projector and scanner can be mounted on one or two stands according to projector's capabilities as discussed in detail in Sect. 2.1.2. Two separate stands or arms give enough flexibility to accommodate various types of projectors. In any case, the stand for the projector must be robust because some of the projectors with adequate parameters can be heavy. However, there exist some portable projectors weighting less than 2 kg as well. A mobile setup with one stand and two separate

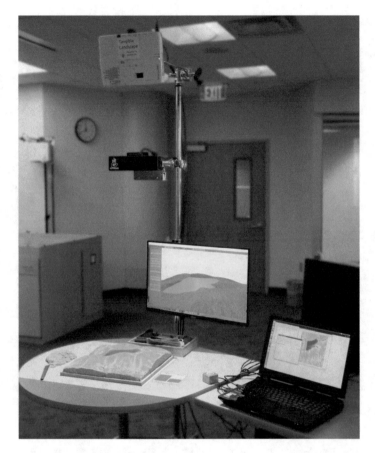

Fig. 2.2 Mobile system setup: projector, scanner, model, laptop and optionally screen

tables is shown in Fig. 2.2, optionally a screen can be attached to the stand. Figure 2.3 shows a similar setup with short throw projector. For permanent setups the projector can be mounted in the ceiling.

2.2 Software

The software for Tangible Landscape consists of two main components:

- Module *r.in.kinect* is a GRASS GIS module which obtains depth and optionally color data from Kinect sensor and processes it into a DEM and RGB rasters. It can run in a loop to continually create new DEMs.

Fig. 2.3 Laboratory system setup with short throw projector and large physical model

- Tangible Landscape Plugin is a GUI plugin of GRASS GIS, which calls *r.in.kinect* and runs selected geospatial analyses on the scanned DEMs as the physical model is being scanned.
- GRASS GIS is the processing engine of Tangible Landscape, which provides all the algorithms and libraries for geocomputation.

Tangible Landscape software is available in the GitHub repository[2] under the GNU GPL license.

2.2.1 GRASS GIS

GRASS GIS (Neteler and Mitasova 2008) is a general purpose cross-platform, open-source geographic information system with raster, vector, 3D raster and image

---

[2]https://github.com/tangible-landscape/grass-tangible-landscape.

Table 2.1 Naming conventions for GRASS GIS modules with examples

Name	Data type	Examples
g.*	General data management	g.list, g.remove, g.manual
r.*	Raster data	r.neighbors, r.viewshed, r.cost
r3.*	3D raster data	r3.colors, r3.to.rast, r3.cross.rast
v.*	Vector data	v.net, v.surf.rst, v.generalize
db.*	Attribute data	db.tables, db.select, db.dropcolumn
t.*	Temporal data	t.register, t.rast.aggregate, t.vect.extract
i.*	Imagery data	i.segment, i.maxlik, i.pca

processing capabilities. It includes more than 400 modules for managing and analyzing geographical data and many more user contributed modules available in the add-on repository. GRASS GIS modules can be run using a command-line interface (CLI) or a native graphical user interface (GUI) called wxGUI which offers a seamless combination of GUI and CLI native to the operating system.

Modules are organized based on the data type they handle and they follow naming conventions explained in Table 2.1. Each module has a set of defined options and flags which are used to specify inputs, outputs, or different module settings. Most core modules using GRASS GIS C library are written in C for performance and portability reasons. Other modules and user scripts are written in Python using the Python Scripting Library which provides a Python interface to GRASS GIS modules.

GRASS GIS software can be downloaded freely from the main GRASS project website.[3] The download web page offers easy to install binary packages for GNU/Linux, Mac OS X, and Microsoft Windows as well as the source code. The GRASS GIS website also provides additional documentation including manual pages, tutorials, information about the externally developed modules (add-ons) and various publications. Support for developers and users is provided by several mailing lists. The following tutorial provides a quick introduction to GRASS GIS: http://grass.osgeo.org/grass72/manuals/helptext.html.

2.2.2   GRASS GIS Python API

GRASS GIS provides several Python application programming interfaces (API) to access different functionalities and accommodate many different use cases. Tangible Landscape uses the Python Scripting Library to easily build new workflows by chaining together existing GRASS GIS modules. We will show an example how to use this library to automate tasks and build new functionality.

---

[3]http://grass.osgeo.org.

GRASS GIS modules are called with the following command syntax:

r.colors   -e   raster=elev_state_500m   color=elevation
‿‿‿‿‿‿   ‿‿   ‿‿‿‿‿‿   ‿‿‿‿‿‿‿‿‿‿‿‿‿‿‿‿
module name flag   option name   option value

This command assigns a predefined color ramp elevation to the raster map
elev_state_500m. It can be executed in a GRASS GIS session from a terminal
or from a GRASS GIS GUI command console. The same command looks very
similar when written using the Python Scripting Library—it is just adjusted to
Python syntax. To run it we have to be in a GRASS GIS session and must first import
the necessary library.[4] We will use function run_command to call individual
modules:

```
import grass.script as gscript
gscript.run_command('r.colors', flags='e',
    raster='elev_state_500m', color='elevation')
```

Besides run_command, other important functions are read_command for reading the raw module output, parse_command for reading module output already
parsed into a Python dictionary, and write_command for passing strings from
standard input. All GRASS GIS modules can be called using these functions.
However, for some commonly used modules the library conveniently provides
wrappers simplifying the syntax, such as the mapcalc function, a wrapper for the
r.mapcalc module.

The following code snippet is a complete Python script which provides a more
complex example using these functions. Here, we extract low-lying areas from a
DEM where the elevation z is lower than the difference $z_{mean} - z_{stddev}$:

```
import grass.script as gscript

def main():
    input_raster = 'elevation'
    output_raster = 'low_areas'
    stats = gscript.parse_command('r.univar', map=input_raster,
        flags='g')
    mean = float(stats['mean'])
    stddev = float(stats['stddev'])
    low = mean - stddev
    gscript.mapcalc('{out} = {inp} <
        {lim}'.format(out=output_raster, inp=input_raster,
        lim=low))

if __name__ == "__main__":
    main()
```

More information on writing Python scripts in GRASS GIS can be found in the
online Python Scripting Library documentation.[5]

---

[4] All further Python code snippets assume the library is already imported to avoid code duplication.

[5] Python Scripting Library documentation: http://grass.osgeo.org/grass72/manuals/libpython/
script_intro.html.

2.2.3   Scanning Module r.in.kinect

GRASS GIS add-on *r.in.kinect* processes raw data from Kinect sensor and processes
it through a series of transformations and filtering procedures to obtain a raster or
vector representation of the physical model. Add-on *r.in.kinect* is written in C++
and uses several open source libraries:

- *libfreenect2* library[6] for access to the color and depth image streams from Kinect
  (Xiang et al. 2016),
- *Point Cloud Library* (PCL)[7] for 3D image and point cloud processing; contains
  numerous algorithms for filtering, segmentation and surface reconstruction (Rusu
  and Cousins 2011),
- *GRASS GIS* library for writing raster and vector data and interpolation.

Using these libraries *r.in.kinect* performs the following basic steps with each
new scan:

1. acquiring the scan as a point cloud,
2. correcting tilting of the scanner through 3D rotation of the point cloud,
3. extracting only relevant points from the point cloud (filtering points outside the
   model area),
4. filtering point cloud to remove outliers,
5. smoothing the point cloud to reduce noise in the data,
6. georeferencing (horizontal rotation, horizontal and vertical scaling, translation)
   the scanned data to known geographic coordinates on the edges of the model,
7. reconstructing a raster digital elevation model using binning or interpolation
   techniques.

Apart from processing the scans, *r.in.kinect* also performs the initial calibration
in two steps. In the first step it estimates the tilting of the sensor in respect to the
table with the model, and in the second step it detects the position of the model on
the table to estimate the cropping extent. After the calibration, the conversion from
the scanned point cloud to a DEM is an automated process; the process' details,
and assumptions that have to be met to produce good results, are described in the
following paragraphs.

Calibrating Scanning Angle During scanning the scanner axis should be oriented
precisely perpendicular to the table with the model in order to avoid a tilted scan.
Even a small deviation of 1 degree can cause centimeters of height difference
depending on the size of the model (see Fig. 2.4a). Therefore, we first calibrate
the system by scanning the empty table and computing the angular deviation. We
then use this information to automatically rotate each scanned point cloud. We
use plane segmentation algorithm (Rusu and Cousins 2011) to extract the part of

---

[6]github.com/OpenKinect/libfreenect2.

[7]github.com/PointCloudLibrary/pcl.

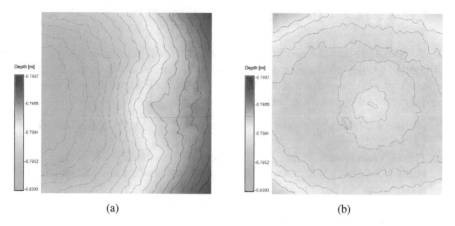

(a)   (b)

Fig. 2.4 Even slight angular deviation from the vertical scanner axis will cause a tilted scan as seen in (**a**) on a scan of a flat plane with a small angular deviation of 1.7°. A calibration based on the initial scan of a flat plane can compensate for such angular deviations resulting in a horizontally aligned scan as shown in (**b**) although a small radial distortion will still be present

the scanned point cloud representing the table and compute the parameters of the resulting plane. We compute the angular deviation between the unit vector along the scanner axis and the vector perpendicular to the plane. We then use Rodrigues' rotation formula (Wikipedia 2015) to derive a rotation matrix to rotate the point cloud by this angular deviation around a vector given as a cross product of the two vectors. After calibration, the derived rotation matrix is used for rotating every scan resulting in perfectly horizontal scan, see Fig. 2.4b. For more precise results, we can also minimize the radial distortion of the scanner visible in Fig. 2.4b. The simplest way to do this is to acquire a scan of a flat surface with the dimensions of the physical model and then subtract it from the raster representation of the model. If the scanner is moved the calibration must be repeated.

Calibrating Model Size and Position Since the scanner captures objects around the physical model, we need to find the boundaries of the model and crop the point cloud by this 3D bounding box in order to georeference it. We first automatically estimate the distance of the scanner from the table where the model will be placed by using plane segmentation algorithm. This value represents the bottom part of the bounding box and allows us to filter the points representing the table (Fig. 2.7a). The horizontal extent of the model is then estimated by identifying the physical model from the scan using Euclidean Cluster Extraction (Rusu and Cousins 2011), and obtaining a 2D bounding box of the identified cluster. Each scanned point cloud is then trimmed using this bounding box, which is slightly enlarged in order to accommodate cases when the model is accidentally moved during scanning. The coordinates of the bounding box can also be specified manually in case the calibration is not suitable for certain models.

Sometimes the edges of physical models are uneven because of sand falling off the sides of the model, which can cause slightly incorrect georeferencing. In those cases additional trimming of the edges may be desirable. This can be successfully done automatically when the model is rectangular, since we can compute a frequency distribution of scanned points of the model in x and y dimensions, and then trim the areas that do not contain enough points with a given threshold from each side of the model (Fig. 2.7c).

Georeferencing   Georeferencing the scanned model is an important step when we need to combine it with our geographic data and to ensure that any geospatial analyses are performed on a DEM with real-world dimensions. We need to designate the DEM raster map that the model represents and specify the vertical exaggeration of the physical model to scale the elevation values properly. The scale factors S_x, S_y and S_z are computed:

$$S_x = \frac{X_{east} - X_{west}}{x_{max} - x_{min}}, \quad S_y = \frac{Y_{north} - Y_{south}}{y_{max} - y_{min}}, \quad S_z = \frac{(S_x + S_y)/2}{e} \qquad (2.1)$$

where X, Y are DEM (real-world) coordinates, x, y are coordinates on the physical model and e is the specified vertical exaggeration. The vertical scale of the model typically differs from the horizontal because we vertically exaggerate the physical model to enhance our perception of the landscape and simplify our interaction with the model. Since Tangible Landscape users usually interact with the physical model from the side opposite to the scanner we have to rotate the scan by 180 degrees in the z-axis. We can georeference the point cloud with the following equation:

$$\begin{bmatrix} X \\ Y \\ Z \end{bmatrix} = \begin{bmatrix} S_x & 0 & 0 \\ 0 & S_y & 0 \\ 0 & 0 & S_z \end{bmatrix} \cdot \begin{bmatrix} \cos\alpha & -\sin\alpha & 0 \\ \sin\alpha & \cos\alpha & 0 \\ 0 & 0 & 1 \end{bmatrix} \cdot \begin{bmatrix} x \\ y \\ z \end{bmatrix} + \begin{bmatrix} t_x \\ t_y \\ t_z \end{bmatrix} \qquad (2.2)$$

which rotates the points around the z axes by angle α in the counterclockwise direction, scales to real-world dimensions, and translates the points by adding t_x and t_y computed so that the lower left corner of the model matches the south-west corner of the DEM. The vertical translation t_z is then similarly computed to match the lowest point of the model and the minimum height of the DEM.

DEM Processing   One of the challenges of DEM reconstruction specific to Kinect is the high noise present in each scan. Certain points of the point cloud are marked as invalid and therefore simple to filter out, others called *flying pixels* (Sarbolandi et al. 2015) can be removed using neighborhood statistics filter, which identifies outliers based on their distance to their k-nearest neighbors (Rusu and Cousins 2011). Further, by applying Moving Least Squares (MLS) surface reconstruction method (Alexa et al. 2003), the point cloud can be resampled and smoothed. The smoothing may vary depending on the application, for example, high smoothing is desired for water flow analysis, but less suitable for solar analysis in urban context.

The point spacing, which determines ideal resolution of the reconstructed DEM, depends on the distance from the scanner, however for most cases 2–3 mm as cell resolution is a suitable value. Note that this resolution is later scaled after georeferencing based on the model scale. Based on the height h of Kinect above the model and Kinect's specifications, the actual point spacing a can be computed:

$$a = h \cdot \tan \left(\frac{70.6°}{512} \right), \qquad h_{min} = 0.5\,\text{m}$$

So for example for height 0.7 m above the model, the point spacing is 1.7 mm.

For creating the DEM from the point cloud we allow users to select either binning or interpolation (see Sect. 6.1). To avoid large number of empty cells, resolution value for binning must be slightly larger than the points spacing value. Interpolation is generally slower, however ensures no empty cells, smoother surface, and allows more flexibility in choosing resolution value. Another approach to improve the DEM quality is to integrate more than one scan into the DEM; this is especially useful when the model has smaller, but important features, which need to be captured (for example markers described in Chap. 4). However, increasing number of points necessarily increases the processing time.

Color Processing Since Kinect provides also color information, we can obtain RGB values together with the depth values. Module *r.in.kinect* then writes red, green and blue components as separate raster layers using binning method. See Sect. 4.4 for examples of using color information to define areas of certain properties. Alternatively, color can be used for tracking laser pointer as the brightest point on the model, which can be used for drawing points, lines or areas. Kinect unfortunately does not allow any control of exposure, often resulting in underexposed or overexposed images. To obtain good results, it is often necessary to manipulate ambient light, brightness of the projector, or the color of the surface on which the physical model lies.

Scanning Speed and Accuracy Tangible Landscape's speed depends upon the size of the model, the point cloud processing methods, and the analyses chosen. Table 2.2 compares approximate times for different point cloud processing methods (binning and interpolation) and analyses (e.g., simulated water flow) for small and medium sized models.[8] If a user interacts with Tangible Landscape immediately after a scan has been captured, then they will have to wait for that scan to be processed before their change will be processed potentially doubling the total processing time.

Figure 2.5 shows an accuracy assessment of scanning, where we compared the difference between original digital elevation model and the scanned and interpolated elevation of a digitally fabricated physical model of the same landscape. Higher differences are to be expected in areas of higher slopes and sharp changes in topography.

---

[8]Benchmarks were performed using a System76 Oryx Pro with i7-6700HQ processor, 16 GB DDR4 RAM, M.2 SSD storage (540 MB/s read, 520 MB/s write), NVIDIA GeForce GTX 1060 Ubuntu 16.04 LTS (64-bit), GRASS GIS 7.2, and Tangible Landscape 2c1ede9.

Table 2.2 Scanning speed for different model sizes

Size, Process	Small	Medium
Physical size	23.5 cm × 23.5 cm	34 cm × 34 cm
Cells	13,456	26,235
Binning	0.51 s	0.71 s
Interpolation	0.74 s	0.97 s
Water flow	0.29 s	1.05 s
Contours	0.05 s	0.06 s
Difference	0.04 s	0.04 s
Landforms	0.03 s	0.08 s

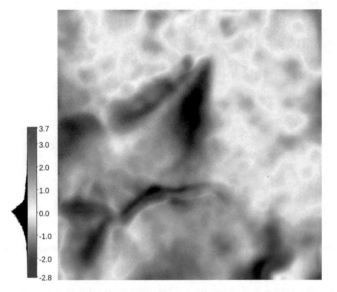

Fig. 2.5 Accuracy assessment: the difference between original digital elevation model and the scanned elevation of a digitally fabricated physical model of the same landscape. The mean difference is -0.02 ± 0.7 mm and the interquartile range is 0.7 mm. The scanned elevation is higher than the original digital elevation model in blue areas and lower in red areas. Legend values are in millimeters

2.2.4   Tangible Landscape Plugin for GRASS GIS

The Tangible Landscape plugin connects the scanning component with GRASS GIS and automates the loop of scanning, importing scans, and geoprocessing in the GRASS GIS environment. It has a graphical user interface which allows the adjustment of the different processing parameters that are necessary to properly georeference and extract the model (Fig. 2.6).

Plugin dialog can be opened from GRASS GIS command console by typing:

```
g.gui.tangible
```

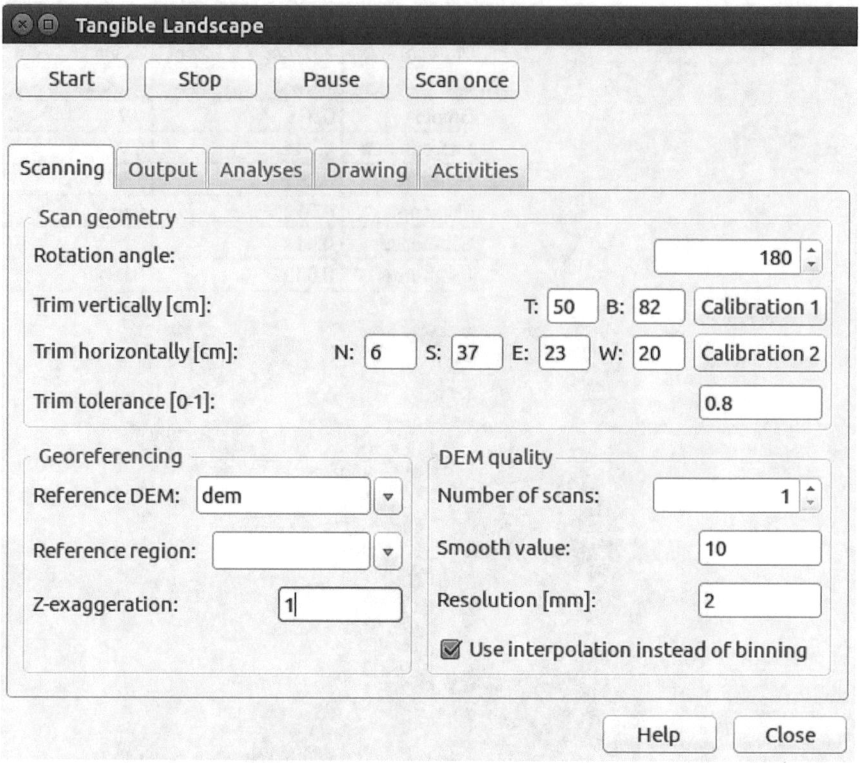

Fig. 2.6 Tangible Landscape plugin dialog (appearance depends on operating system)

The first two buttons of the topmost button group start and stop the continuous scanning process. Since stopping and restarting the process takes up to several seconds and switches off the scanner, it is often advantageous to use the *Pause* button to just temporarily stop the processing of the scan while keeping the scanner running and at the same temperature preventing measurement drift. *Scan once* launches the scanning process in a single-scan mode, which is useful when continuous scanning is not needed.

Scanning Tab There are several tabs below the buttons to parametrize different aspects of scanning and analysis. Parameters in scanning tab are grouped based on their effect on scan geometry, georeferencing and reconstructed DEM quality, and they reflect the parameters of *r.in.kinect* module. Scan geometry refers to the relative position of the scanner and the physical model. It includes the rotation along z-axis, which is used when the scanner is oriented differently than the model, and the coordinates of the 3D bounding box (relative to the scanner position) limiting the scanned point cloud (Fig. 2.7a, b). These values can be in most cases calibrated automatically. The *Trim tolerance* value can optionally be used to automatically find best edges of a rectangular model for each scan, which makes georeferencing of the model more precise in case of lying sand or other objects around the model or hands reaching over the model.

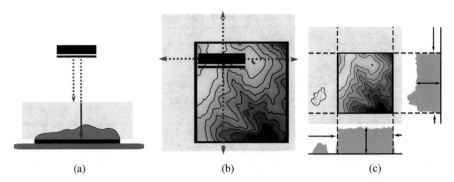

Fig. 2.7 Finding the boundaries of the physical model: (**a**) setting vertical limits for scanning the physical model (distances from the scanner) and (**b**) setting horizontal limits (horizontal distances from the scanner in north, south, east, and west direction). Automated trimming of edges (**c**) of a rectangular model finds exact edges of the model to make georeferencing less sensitive to objects and hands around the model

Georeferencing options allow us to determine the geographical location and extent the model represents by providing a reference DEM (or alternatively a named GRASS region). Since we often vertically exaggerate the physical model to highlight terrain features, we can specify the exaggeration factor, which is used subsequently for correcting the scanned digital DEM, in order to run simulations on landscapes without exaggeration.

There are several parameters which influence the quality of the reconstructed DEM, including the number of scans integrated into the final DEM, level of smoothing, reconstruction method (binning or interpolation), and DEM resolution. Certain applications, such as water flow modeling, can benefit from highly smoothed, interpolated surface, while change detection typically works well even with rougher binned surface, which is faster to reconstruct. We can change all these parameters during scanning to see the resulting DEM and select the optimal combination of parameters.

Output Tab In *Output* tab we can specify the names of the scanned DEM (the default is *scan*) and the prefix for the three color raster layers (red, green and blue channels) if we need them. Besides the R, G, B raster layers, GRASS imagery group named as the specified prefix is created to allow using the color outputs easily with GRASS imagery modules. For cases when a point cloud is more advantageous representation, a PLY file can be automatically exported. In all cases, each new scan overwrites the previous files.

Analyses Tab This section allows to customize the geospatial analyses running for each scan. Here you can specify the path to your Python file where you call all your analyses. The button *Create new file with predefined analyses* creates a new Python file which already contains examples of analyses that can be used as they are, or as a starting point for more complex workflows. The file contains several functions

Table 2.3 Available parameters of functions ran for each scan

Function parameter	Purpose
`real_elev`	Name of the reference DEM
`scanned_elev`	Name of the scanned DEM
`scanned_color`	Basename for the RGB rasters, imagery group name
`blender_path`	Directory monitored by Blender (Sect. 5.3)
`zexag`	Currently set vertical exaggeration
`eventHandler`	Used for updating dashboards (Sect. 14.1)
`env`	Environment for running modules[a]

[a]Defines computational region matching the scan, and other environment variables controlling overwriting outputs and verbosity level

(initially commented out), where each function represents separate analysis, for example:

```
def run_contours(scanned_elev, env, **kwargs):
    analyses.contours(scanned_elev=scanned_elev,
        new='contours_scanned', env=env, step=2)
```

Uncommenting this function activates contour computation with contour interval of 2 vertical units (the suitable interval depends on your reference DEM). Once Tangible Landscape runs, after each scanning cycle a vector map `contours_scanned` (the name can be changed) is created. You can add that vector map to the Map Display. During scanning you can add new functions or change the code of the functions, for example to adjust the contour interval, and once the file is saved the change is adopted. This enables us to dynamically develop and test new workflows during scanning. Each function is independently run for each scan. The function name must start with prefix `run_` and has several parameters, where some of them don't have to be listed as they are useful only for specific workflows (Table 2.3):

```
def run_myanalysis(real_elev, scanned_elev, scanned_color,
    blender_path, zexag, eventHandler, env, **kwargs):
    # do computation
```

The plugin provides a library containing often used analyses and workflows which can be readily used and combined with other GRASS GIS functionality. The analyses can be scripted using GRASS GIS Python API (see Sect. 2.2.2). Additional examples of analyses can be found on GRASS GIS Wiki.[9]

Drawing Tab Tangible Landscape allows using laser pointer to draw points, lines and polygons on the model. This tabs allows you to activate this functionality and select which vector type to draw. While drawing, *r.in.kinect* looks in a loop for

---

[9]https://grasswiki.osgeo.org/wiki/Using_GRASS_GIS_through_Python_and_tangible_interfaces_(workshop_at_FOSS4G_NA_2016)#Tangible_Landscape.

the brightest point (sum of R, G, and B values) above certain threshold and keeps recording it until the drawing is ended. Then the vector layer is created and can be used in further workflows. This method is fairly sensitive to the overall brightness and ambient light, therefore the threshold value needs to be tested and adjusted whenever the conditions change.

Activities Tab When using Tangible Landscape as a teaching tool, it is advantageous to go through several exercises and demonstrations of geospatial concepts. In order to quickly switch between these different analyses, we can define separate activities, where each activity is defined by Python file with analyses, map layers loaded in the beginning of each activity, scanning parameters, optional simple dashboard and slides accompanying the activity. These components are specified in JSON configuration file, which is described on Tangible Landscape wiki.[10]

2.2.5 Tangible Landscape Plugin Installation

Since Tangible Landscape is currently not packaged, we describe here the general steps for its compilation on any platform. Some experience with compilation is advantageous.

1. Install (either compile or use packaged) dependencies, namely Point Cloud Library v1.8, libfreenect2 v0.2, GRASS GIS v7.4 using each project's installation instructions.
2. Download and compile *r.in.kinect*.
3. Install Tangible Landscape plugin using GRASS module *g.extension*, alternatively download and compile it yourself.

In the Tangible Landscape repository we provide installation script for GNU/Linux distribution Ubuntu, which is the preferred platform for Tangible Landscape. The installation is then simplified into a few lines:

```
mkdir dev && cd dev
wget https://raw.githubusercontent.com/tangible-landscape/grass\
-tangible-landscape/master/install.sh
sudo sh install.sh
```

The script would be similar for Mac OS, however we recommend using homebrew package manager[11] for installing dependencies. Compilation on Windows platform requires extensive experience and is not recommended.

---

[10]https://github.com/tangible-landscape/grass-tangible-landscape/wiki/Working-with-Activities.
[11]https://brew.sh/.

References

Alexa, M., Behr, J., Cohen-Or, D., Fleishman, S., Levin, D., & Silva, C. T. (2003). Computing and rendering point set surfaces. *IEEE Transactions on visualization and computer graphics, 9*(1), 3–15.

Andersen, M., Jensen, T., Lisouski, P., Mortensen, A., Hansen, M., Gregersen, T., & Ahrendt, P. (2012). Kinect depth sensor evaluation for computer vision applications. Technical Report, Aarhus University, Department of Engineering. Denmark.

Gonzalez-Jorge, H., Riveiro, B., Vazquez-Fernandez, E., Martínez-Sánchez, J., & Arias, P. (2013). Metrological evaluation of Microsoft Kinect and Asus Xtion sensors. *Measurement, 46*(6), 1800–1806.

Lachat, E., Macher, H., Mittet, M.-A., Landes, T., & Grussenmeyer, P. (2015). First experiences with Kinect V2 sensor for close range 3D modelling. *ISPRS - International Archives of the Photogrammetry, Remote Sensing and Spatial Information Sciences, XL-5/W4*(February), 93–100.

Mutto, C., Zanuttigh, P., & Cortelazzo, G. (2012). Introduction. In *Time-of-flight cameras and microsoft kinect (TM)*. Springer briefs in electrical and computer engineering (pp. 1–15). Boston, MA: Springer US.

Neteler, M., & Mitasova, H. (2008). *Open source GIS: A GRASS GIS approach* (3rd ed.). New York: Springer.

Rusu, R. B., & Cousins, S. (2011). 3D is here: Point Cloud Library (PCL). In *IEEE International Conference on Robotics and Automation (ICRA)*, Shanghai, China.

Sarbolandi, H., Lefloch, D., & Kolb, A. (2015). Kinect range sensing: Structured-light versus time-of-flight kinect. *Computer Vision and Image Understanding, 139*, 1–20.

Wasenmüller, O., & Stricker, D. (2017). Comparison of kinect V1 and V2 depth images in terms of accuracy and precision. In C.-S. Chen, J. Lu, & K.-K. Ma (Eds.), *Computer Vision – ACCV 2016 Workshops: ACCV 2016 International Workshops*, Taipei, Taiwan, November 20–24, 2016. Revised Selected Papers, Part II (pp. 34–45). Cham: Springer International Publishing.

Wikipedia (2015). Rodrigues' rotation formula — wikipedia, the free encyclopedia [online]. Accessed 13.08.2015. https://en.wikipedia.org/w/index.php?title=Rodrigues%27_rotation_formula&oldid=671556479

Xiang, L., Echtler, F., Kerl, C., Wiedemeyer, T., Lars, hanyazou, Gordon, R., Facioni, F., laborer2008, Wareham, R., Goldhoorn, M., alberth, gaborpapp, Fuchs, S., jmtatsch, Blake, J., Federico, Jungkurth, H., Mingze, Y., vinouz, Coleman, D., Burns, B., Rawat, R., Mokhov, S., Reynolds, P., Viau, P., Fraissinet-Tachet, M., Ludique, Billingham, J., & Alistair (2016). libfreenect2: Release 0.2. https://doi.org/10.5281/zenodo.594510

Chapter 3
Building Physical 3D Models

Tangible Landscape works with many types of physical 3D models. When used to sculpt topography the physical model should be built of a malleable material such as sand or clay so that users can easily deform the surface. When used for object recognition the physical model can be built of a rigid material such as a wood product, foam, plastic, or resin. When both modes of interaction are combined the physical model should use malleable materials for the base and rigid materials for the objects. These models can be built by hand or digitally fabricated using 3D printing or computer numerically controlled (CNC) manufacturing (Fig. 3.1). Tangible Landscape's difference analytic can be used as an aid for hand-making models. The final model should be opaque, have a light color, and have a matte finish so that the projected image is crisp and vivid. The final model should be opaque, have a light color, and have a matte finish so that the projected image is crisp and vivid, since transparent materials such as acrylic cannot be 3D scanned. Some 3D printing and casting materials like resin may appear opaque, but have translucent properties—this will diffuse the projection. If we desire a very crisp and vivid image on a rigid model made of wood products or resins we recommend painting the model white. In this chapter we discuss different types of physical models and explain how to fabricate them.

3.1 Handmade Models

Rigid physical models can be handmade by hand-cutting contours and malleable physical models can be handmade by sculpting sand or clay.

Contour Models Contour models can be precise if they are finely cut, but they are inaccurate as they depict abstract, stepped landscapes that are discrete rather than continuous. They are also very legible—one can easily count the contours and read the height—but again they represent abstracted landscapes. While one can easily

© The Author(s) 2018
A. Petrasova et al., *Tangible Modeling with Open Source GIS*,
https://doi.org/10.1007/978-3-319-89303-7_3

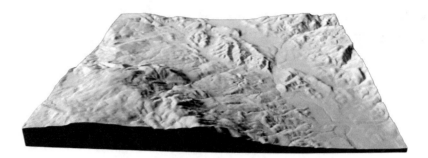

Fig. 3.1 A CNC routed model of the Sonoma Valley, California

read the contours and then calculate slope, it is hard to intuitively visualize the slope. Furthermore this abstract, discontinuous representation may obfuscate the relationship between form and process. Furthermore they are time consuming and complicated to construct especially for complex topographies. Hand cutting contour models can be dangerous as the knife may slip or jump.

Contour models can be made out of stacked boards of paperboard, cardboard, foam, or a soft wood like balsa or basswood. The thickness of the sheets should correspond to the vertical interval between contours at the desired map scale. To build a contour model by hand start by plotting a contour map at the desired map scale. Then spray the plotted paper map with a spray mount adhesive, and stick the plotted map onto a board. As an alternative to plotting and mounting, we can instead project the contour map. Cut out the lowest contour level with a precision knife such an x-acto. Glue this contour level onto the base or the level below. Repeat this process until all of the contour levels have been cut out and stacked together.

Hand Sculpted Models Hand sculpted sand and clay models are natural, intuitive, and even fun to make, but are imprecise and hard to read quantitatively. When sculpting by hand we not only see, but also feel the 3-dimensional form, using our highly developed kinaesthetic intelligence to better understand the space. While tools can help us to cut sharp edges, shape certain forms, and smooth surfaces, sculpting with our hands alone lets us feel the shape of the topography and get an intuitive understanding of its form undistorted by perspective and depth perception. Because these models are continuous they could—if sculpted well—accurately represent the topography, but as they lack discrete intervals it is challenging to quantitatively judge distances and heights. Because it is so natural and easy to make, but challenging to read quantitatively freeform hand modeling is better suited to ideation than detailed design or presentation.

While hand sculpted models have traditionally been made out of sand or clay, we recommend using polymer enriched sand. Clay holds form well, but can be sticky and hard to work with. Sand, though easy to move, does not hold a shape well—sculpt too steep a slope and it will slump, the grains avalanching down. Polymeric sand is easy to work with and holds together well without sticking to

Fig. 3.2 Sculpting a polymeric sand model with the aid of a projected elevation map and contours

our hands. It is not only easy to sculpt, but also easy to cast into precise forms. We will discuss methods for casting sand in digitally fabricated molds in Sect. 3.3. We use the product DeltaSand, which is also marketed as Kinetic Sand. It is made by mixing sand with a low-viscosity polymer and a cross-linking agent, resulting in a sand with the plasticity, moldability, adhesiveness, and longevity required for sculpting (Modell and Thuresson 2009).

Tangible Landscape can be also be used as an aid for hand sculpting models. Users can project the digital elevation model and contours that they want to build over their polymeric sand model as a static guide for sculpting (Fig. 3.2). They can also use Tangible Landscape's DEM differencing as a real-time guide about where to add or remove sand. This analysis computes the difference between the target digital elevation model and the scanned model that has been sculpted (Fig. 3.3). See Sect. 6.2.2 for a detailed description of DEM differencing.

3.2   Digitally Fabricated Models

Complex physical models for Tangible Landscape can be digitally fabricated with CNC routing or 3D printing. Digitally fabricated models are rigid and precise— while they can not easily be sculpted, they are ideal presentation models, base models for object recognition, and molds for casting malleable models. CNC

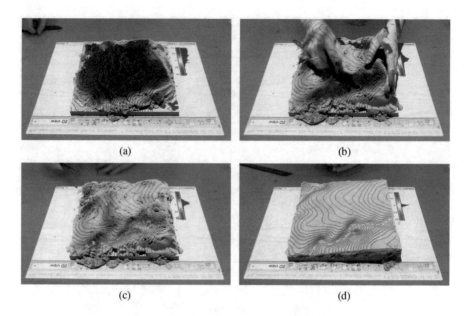

(a)   (b)

(c)   (d)

Fig. 3.3 Hand sculpting a terrain model from polymer enriched sand with the aid of Tangible Landscape's DEM differencing where blue is too low and red is too high: (**a**) unsculpted sand showing major differences, (**b**) sculpting the sand to reduce differences, (**c**) intermediate model with moderate differences, (**d**) and final terrain model with minimal differences

machines such as laser cutters, routers, and 3D printers require digital instructions, a controller to decode the instructions, and a machining tool. Machining processes include subtraction with routers, addition with 3D printers, and deformation and molding with vacuum formers (Schodek et al. 2004; chap. 13). In this section we will explain how to prepare GIS data for digital fabrication and discuss various methods of digital fabrication. Section 3.4 describes workflows for creating models using these methods.

3.2.1   Digital Models

Terrain and other continuous, 3D geographic data can be digitally represented as 3D points, 2.5D rasters, 3D rasters, contours, meshes, or surfaces. In GRASS GIS our workflows for landscape modeling and analysis are raster based. 3D printers and CNC routers, however, require vector data, typically meshes. After acquiring elevation data we import them into GRASS GIS and if necessary we interpolate a raster digital elevation model (DEM). This raster will later be used as the reference data for Tangible Landscape. Next, we export the raster as a point cloud and import it into a 3D modeling program to compute a surface, mesh, or solid from the point cloud. Finally, we generate a toolpath from the surface or mesh for the 3D printer or CNC router.

Digital Elevation Models The first step in modeling a landscape is to acquire data. Entire landscapes can be precisely 3D scanned as point clouds with airborne lidar. As pulses of light are emitted from the aircraft they hit and reflect back from vegetation, structures, and topography. When a pulse of light hits a tree part of the light is reflected and recorded as the first return, while the rest penetrates the outer canopy. The residual pulse, recorded as a series of returns, reflects off of leaves, branches, shrubs, and the ground. Lidar data can be used to compute very precise, high resolution models. By filtering lidar returns and classes we can extract points for the bare earth topography and interpolate a raster DEM or we can extract points with all of the structures and vegetation and interpolate a raster digital surface model (DSM). Lidar point clouds or ready to use raster DEMs are available for many locations and are well suited for building topographic models (see Appendix A.2 for links to lidar data and DEM repositories).

DEMs—raster maps of bare earth topography—and DSMs—raster maps of topography with structures and vegetation—are commonly used in GIS to represent and analyze terrain as they can easily be mathematically transformed. For example topographic parameters such as slope, aspect, and curvature can be computed from the partial derivatives of a function approximating a DEM or DSM (see Chap. 6).

Lidar, especially when filtered by classes or returns, can produce scattered, spatially heterogeneous data points that are challenging to accurately interpolate as terrain surfaces. Furthermore, other sources of elevation data such as digitized contours and surveying data can be clustered and have a highly spatially heterogeneous distribution. The regularized spline with tension (RST) interpolation function, implemented as the *v.surf.rst* module in GRASS GIS, can be used to accurately model terrain surfaces from clustered and heterogeneous data such as lidar with some experimentation and tuning (see Sect. 6.1.2).

Contour data are challenging to interpolate because the data are spatially heterogenous—they are often very dense along the contour lines with large unsampled areas in regions with flat topography. If we acquire digital contour data we can either linearly interpolate between contours with *r.surf.contour* or thin the points along contours using *v.generalize*, convert them to a point representation and then interpolate them using a function like RST with the module *v.surf.rst* (The GRASS GIS Community 2015).

Meshes Terrain can also be modeled in 3D vector representations such as polygon meshes or mathematically defined surfaces. In a mesh a vector network of polygons such as triangles or quadrilaterals forms a shape. Terrain meshes are typically triangulated irregular networks (TINs) that connect the data points from which they are computed. Triangulation was used to manually interpolate contours from surveyed spot elevations before the advent of computers. Now TINs are typically computed using the Delaunay triangulation algorithm which maximizes the smallest angle of every triangle in order to minimize the occurrence of very thin triangles. Since TINs connect the data points the input data is preserved. When meshes are low resolution, i.e. computed from small datasets, they reveal their polygonal structure with angled planes where there should be curved surfaces. When meshes are high resolution, i.e. computed from large datasets, they can be very detailed and represent highly complex forms. Even a high resolution mesh, however, is inaccurate when used to represent curved slopes as planar faces. Meshes are very useful for modeling structures and engineered topography with discontinuous slopes.

NURBS Non-uniform rational B-splines (NURBS) are parametric approximation curves and surfaces defined by a series of polynomials. The curves are basis splines defined by a knot vector, weighted control points, and the curve's order or degree (Piegl and Tiller 1995). A NURBS surface is the tensor product of two NURBS curves (Martisek and Prochazkova 2010). Since NURBS surfaces are mathematically defined they can precisely describe a continuous geometry.

Processed lidar data and rasters can be exported as point clouds from GIS and then imported into a 3D modeling program for meshing or surface generation. Toolpaths for CNC routing can be generated for a mesh or NURBS surface once it has been scaled. 3D printing, however, requires closed, solid volumes so the terrain surface should be offset or extruded to give it a thickness. Most 3D printers read the stereolithography (.stl) format, a mesh representation of a solid.

We typically use Rhinoceros for 3D modeling[1] with the RhinoTerrain plugin for terrain modeling[2] and the RhinoCAM plugin for CNC routing.[3] This proprietary 3D modeling program is useful for its interoperability—it can write and read a wide range of formats and can model NURBS, polygon meshes, solids, and point clouds. The RhinoTerrain plugin has modules for importing and exporting geographic data and efficiently computing large terrain meshes. While many computer aided manufacturing (CAM) programs require polygon meshes, the RhinoCAM plugin enables us to generate and simulate CNC toolpaths in Rhinoceros using NURBS, polygon meshes, or solids. The Rhino3DPRINT plugin can be used to prepare models for 3D printing.[4] Alternatives include a proprietary 3D modeling and animation program,[5] or MeshLab, an open source mesh processing program.[6]

3.2.2 Laser Cutting

Laser cutting can be used to efficiently build precise contour models of moderate complexity (Fig. 3.4). Laser cutters are CNC machines that use a laser moving along a gantry in the x and y axes to make 2D cuts. To laser cut a contour model we separate each contour step onto different layers of a 2D vector drawing in a

---

[1] Rhinoceros: http://www.rhino3d.com/.

[2] RhinoTerrain: http://www.rhinoterrain.com/.

[3] RhinoCAM: http://mecsoft.com/rhinocam-software/.

[4] Rhino3DPRINT: http://mecsoft.com/rhino3dprint/.

[5] Maya: http://www.autodesk.com/products/maya/overview.

[6] MeshLab: http://meshlab.sourceforge.net/.

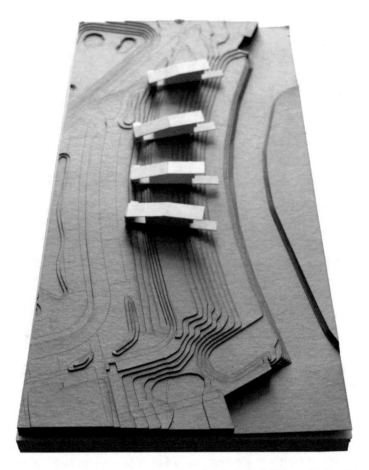

Fig. 3.4 A laser cut chipboard contour model with basswood buildings fabricated by David Koontz and Faustine Pastor

computer-aided design (CAD) format (like .dwg, .dxf, or .ai). We cut each contour step out of a sheet of our material of choice and glue them together. We can also laser etch elevation values, patterns, and line work like road outlines and building footprints onto each layer. While it is easy to precisely cut complex topographies it is still challenging to construct them as we have to lay and glue each piece by hand.

The materials determine the cost. As this is a subtractive method a lot of material is wasted. Contour models cut from thick, low-cost materials like cardboard are relatively inexpensive, but have large contour steps and thus low resolution. Models cut from thin, low-cost materials like chipboard allow for small contour steps at moderate cost, but tend to warp with time especially if too much glue has been applied. When aesthetics and archivability are important museum board or acrylic can be cut to create presentation-quality laser cut contour models. Paper-based media such as chipboard or museum board are not suitable for casting polymeric sand as the polymer will stick and soak into them.

Fig. 3.5 A CNC routed terrain model of the Jockey's Ridge dune complex on the Outer Banks of North Carolina

3.2.3 CNC Routing

CNC routing or milling is a subtractive fabrication process that can be used to precisely manufacture contour models and surface models of great complexity (Fig. 3.5). This type of digital fabrication is a precise, accurate, inexpensive, and scalable way to build terrain models. 3-axis CNC routers and milling machines move a spindle with a machining bit along the x, y, and z axes in a programmed path to carve a shape out of a block of material (Fig. 3.6). Because 3-axis routers can only carve vertically they produce 2.5D surface models on a solid base, rather than full 3D volumetric models. It is possible, however, to manufacture fully 3D volumes with 3-axis routers; after each cut one can reorient the model and then route the sides or bottom of the model (Schodek et al. 2004; chap. 14).

To CNC route a terrain model we first prepare a solid block of our material. We can use materials such as foam, wax, and wood products. We typically use medium density fiberboard (MDF), an engineered wood product with high strength, good dimensional stability, and a fine, uniform grain with minimal voids. Some suppliers manufacture blocks of MDF up to 4-in. thick. If your suppliers only stock 0.25- and 0.5-in. thick sheets of MDF, then you can create a block of custom thickness by cutting the sheets into tiles that match the desired extent of our scale model,

Fig. 3.6 A 3-axis CNC router carving a terrain model

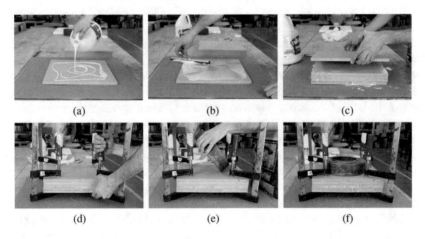

(a)   (b)   (c)

(d)   (e)   (f)

Fig. 3.7 Preparing MDF for CNC routing: (**a**) pour glue onto a layer of MDF, (**b**) spread glue evenly over a layer of MDF, (**c**) stack layers of MDF, (**d**) clamp the stack of MDF, (**e**) add a weight on top of the stack of MDF, and (**f**) wait for the glue to dry

spreading wood glue evenly over each tile and stacking the tiles together, and firmly clamping the stack of tiles together until the glue sets (Fig. 3.7). MDF is relatively inexpensive and very durable, but is heavy, only machines moderately well, and the particulate produced during machining is a serious health hazard. Low density polystyrene foams are inexpensive and lightweight, but machine poorly and are not

Table 3.1 CNC settings: horizontal rough cut with MDF

Bit ∅ (in.)	Speed	Plunge	Approach	Engage	Cut	Retract	Depart.	Step down (in.)	Stepover (%)
0.5	16,000	65	135	205	275	800	800	0.32	48
0.25	16,000	50	75	125	175	800	800	0.25	40

Table 3.2 CNC settings: parallel finish cut with MDF

Bit ∅ (in.)	Speed	Plunge	Approach	Engage	Cut	Retract	Departure	Stepover (%)
0.5	16,000	65	160	230	325	800	800	20
0.25	16,000	50	100	150	225	800	800	20

durable. Typically polystyrene foams are manufactured in colors such as light blue and pink that are not suitable for terrain models. These foams dissolve if spray painted, unless a water-based paint is used or the foam is primed prior to painting. Medium and high density polyurethane foams such as RenShape are designed for high precision machining. These foams are manufactured in sheets and blocks up to 4-in. thick. They are relatively light weight, machine finely—producing very detailed models—but can be very expensive. We often use RenShape 5020 foam board for its weight, machinability, color, and cost.

In a CAD program we prepare a digital model of the geometry we wish to carve. Then we generate a toolpath for this digital model in a CAM program. To make a contour model we use closed, 3D contour curves as our data and then carve with a contour cut. Contour cutting can also be used to carve buildings. To carve a surface model from a mesh or NURBS surface we use parallel cuts. For a surface model we typically start with a horizontal rough cut with a 0.5 in. diameter, carbide bit to remove the bulk of the excess material (Table 3.1). Then we use a parallel finish cut with a 0.25 in. diameter, carbide, ball-nose bit to carve the terrain as a continuous surface (Table 3.2). If we need a more refined presentation-quality model we then make two more parallel finish cuts in alternative cutting directions with a 0.125 in. diameter, carbide, ball-nose bit to smooth the surface and capture more detail. Finally a contour cut along the border can be used to neatly trim and remove the model from the base material. The toolpath is written as a sequence of instructions in a CNC programming language often as a .nc file in G-Code. The CNC machine's controller reads this code and drives the tool along the toolpath carving our model. We have a streamlined workflow using Rhinoceros with the RhinoCAM plugin for both 3D modeling and the generation, simulation, and visualization of toolpaths. We do our CAD and CAM in the same environment—Rhinoceros—so that we can easily edit our models and work with both meshes and NURBS.

Once the model has been routed we may want to finish it by sanding, priming, and painting the model. Sanding the model, a prerequisite for priming and painting, with a fine grit sandpaper produces a smoother surface. After routing and sanding we clean the model with an air hose. Applying several coats of magnetic primer to our model weakly magnetizes it so that magnetized markers stick to the slopes (Fig. 3.8).

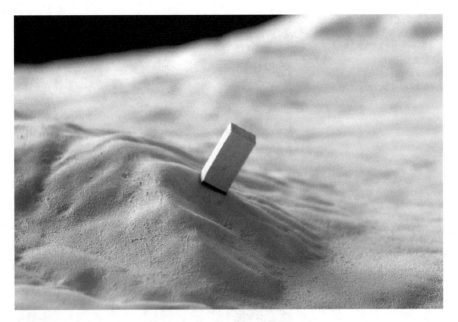

Fig. 3.8 A marker with a magnetic base sticking to CNC-routed terrain model primed with magnetic paint

A magnetized model is useful for Tangible Landscape applications relying on object recognition rather than sculpting (Sect. 4.3). Finally we may want to spray paint our model white to enhance the brightness of projected imagery.

3.2.4   3D Printing

With 3D printing, a type of solid freeform fabrication, we can make a complex, 3D volume in a single run. While the models can be precise, accurate, and highly complex, they are also expensive and small as 3D printers have restrictively small build areas and the materials are relatively expensive. 3D printing is an ideal process for fabricating small, complex models such as buildings or small, high quality presentation models for use with Tangible Landscape (Fig. 3.9a).

While CNC milling and routing are subtractive processes, 3D printing is an additive process in which a model is built up layer by layer. A digital, 3D, solid model is divided into a stack of horizontal layers or slices and a toolpath is computed for each slice. The physical model is then built slice by slice by depositing or hardening material along the toolpath. By dividing the model into cross sections, each as thin as the technology allows, a complex volume can be formed. There are a variety of different 3D printing processes including stereolithography, selective laser sintering, fused deposition modeling, and 3D ink-jet printing each with tradeoffs in build speed, quality, strength, cost, resolution, color, and material (Schodek et al. 2004; chap. 14).

Fig. 3.9 Casting polymeric
sand from 3D printed molds:
(**a**) 3D prints of the terrain
and canopy, (**b**) casting sand
with 3D prints, and (**c**) the
canopy cast in sand

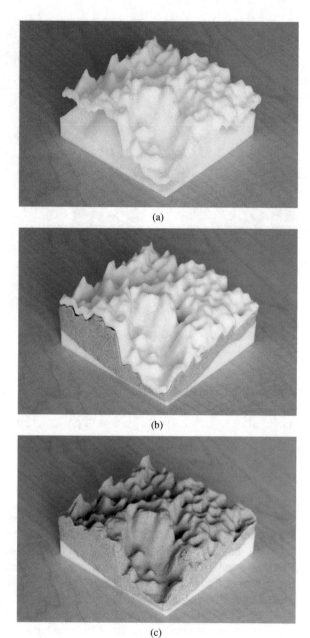

(a)

(b)

(c)

3.3   Molding and Casting

To sculpt with Tangible Landscape we need a malleable model made of a soft, deformable medium. A model made of polymer enriched sand is easy to sculpt, holds its form well, and can be cast into precise forms. CNC routed and 3D printed models can be used as molds for casting polymeric sand into malleable models. The mold should be the inverse of the surface that will be cast. Cast models can precisely represent complex forms that are too challenging to model by hand and can easily be recast.

To cast a terrain model either CNC route the inverse of the terrain (Fig. 3.10) or 3D print the terrain as a volume extruded with enough thickness to survive the casting process (Fig. 3.9). We press the mold into the polymeric sand to cast the form. We have to check the cast and repeat the process if necessary—sand may need to be added or removed in places to get a good cast. Clamps can be used to get strong, even pressure on a CNC routed mold when casting.

Thermoforming or vacuuming forming can be used to quickly make lightweight copies or imprints of a CNC routed terrain model in a thermoplastic (Fig. 3.11). Since they are lightweight thermoformed models are very portable and are ideal for casting polymeric sand while traveling. To thermoform a terrain model we heat a thermoplastic sheet in a vacuum former until it becomes soft. Use the vacuum to pull the hot plastic over a mold deforming the plastic into the desired shape. Once the plastic cools into the cast shape we release the vacuum. We may need to drill small holes through our mold to make a complete vacuum (Schodek et al. 2004; chap. 14).

Fig. 3.10 Cast polymeric sand from CNC routed molds: (**a**) pour polymeric sand onto the base, (**b**) stack the inverse model on top of the sand, (**c**) apply pressure, (**d**) check the cast and remold if necessary, (**e**) trim the excess sand, and (**f**) remove the inverse model

Fig. 3.11 A thermoformed
polystyrene model of part of
Lake Raleigh Woods cast
over a CNC routed mold

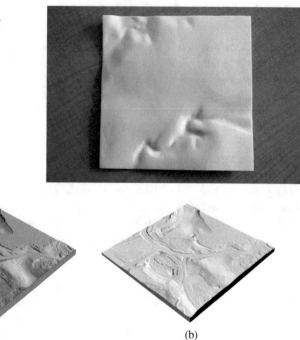

(a)   (b)

Fig. 3.12 3D renderings of terrain models derived from lidar tile 0793_015 using (**a**) GRASS GIS
and (**b**) Rhinoceros with the RhinoTerrain plugin (see Sect. 3.4.4)

3.4 Workflows

This section describes workflows for creating physical models for Tangible Landscape. Our examples use lidar point cloud data for Lake Raleigh Woods on North
Carolina State University's (NCSU) Centennial Campus (Fig. 3.12) acquired during
a 2013 lidar survey by Wake County. The data are provided as a single tile
in the standard LAS format (`tile_0793_015_spm.las`). We processed this
data within a GRASS GIS data set `nc_spm_tl`. All examples in the following
subsections assume running GRASS GIS with this data set; see Appendix A.3 for
information on how to download the data, start GRASS GIS with this location data
set, and perform basic display and other operations.

3.4.1 Selecting a 3D Model Scale

The scale of the physical 3D model depends on the extent of the data, the desired
model size, and the desired scale. The scale may also depend on the material of the
model, the scanning technology, and the resolution of the data. Here we assume that

we know the spatial extent of our data and want our model to be approximately half a meter long on each side. If we do not already know the extent of our data we can determine the values once the data has been imported into GRASS GIS for example using *v.info* or *r.info*.

We can perform this computation in the Python interactive console. First, we enter the spatial extent of the data and approximate the size of the model (in meters):

```
n = 224028.45
s = 223266.45
e = 639319.28
w = 638557.28
desired_model_x = 0.50
```

Then we compute the scale using the equation:

$$s = \frac{d_m}{d_r} \qquad (3.1)$$

where s is the scale of the model, d_m is the distance measured on the model, and d_r is the real-world distance. For convenience we can also compute the scale number using the equation:

$$s_n = \frac{1}{s} \qquad (3.2)$$

where s_n is the scale number of the model and s is the scale of the model.

We then type the following in the Python console:

```
scale = desired_model_x / (n - s)
scale_number = 1. / scale
```

The scale is now $1 : 1524$ and the scale number is 1524. Since we want the scale to be a round number and do not want to add or remove data we need to modify the scale and reverse the computation in the previous step so that we can compute the new size of the model at the selected scale. We select a scale of $1 : 1500$ (a round value close to the previously computed scale). In Python we then write:

```
scale_number = 1500
scale = 1. / scale_number
model_x = (n - s) * scale
```

The resulting model size is 50.8 cm.

So far we have only used the spatial extent along the north-south direction. We skipped the east-west direction because the extent of both the data and the model were squares. In a region with an elongated rectangular shape we might need to determine a suitable scale for each of the horizontal directions and then decide on a compromise.

Next we determine the vertical size of the model and its vertical exaggeration. First, we enter the minimum and maximum elevation in the dataset (called top and bottom in GRASS GIS) and we also include the desired height of the model (in meters). In our example we set the desired height of the model to 0.04 m:

```
t = 109.33
b = 76.54
desired_model_h = 0.04
```

Then we compute the model height without vertical exaggeration and the exaggeration based on the desired model height using the equation:

$$e = \frac{h_e}{h} \tag{3.3}$$

where e is the exaggeration, h_e is the exaggerated height (in our case the desired height of the model) and h is the height of the model according to the scale in case the model would not be exaggerated.

Since we already know the scale of the model from the previous computations we can now compute the exaggeration and model height without vertical exaggeration:

```
model_h = scale * (t - b)
exaggeration = desired_model_h / model_h
```

With the given the inputs the exaggeration is now approximately 1.8. Since we want a round number we set the vertical exaggeration to 2 and compute the actual height of the model using Eq. (3.3):

```
exaggeration = 2
actual_model_h = exaggeration * model_h
```

With the exaggeration set to 2 the model height (or more precisely the maximum height difference of the top surface of the model) will be approximately 4.4 cm. The base of the model will increase the actual model height. Depending on how it is constructed the base may or may not be an integral part of the model. If we set the exaggeration to 1.5 the model height difference would be approximately 3.3 cm, which might be too shallow. At a different scale, however, an exaggeration of 1.5 might be the right choice.

We have used the same scale in both the x and y horizontal directions (1 : 1500). However, in the vertical direction (z) we exaggerated the scale by a factor of 2. As a result the vertical scale (1 : 750) is different than horizontal scale (1 : 1500). Depending on the audience it may be advantageous to note either the horizontal and vertical scales or the scale and exaggeration. When the scale and exaggeration are known the vertical scale can be computed using the equation:

$$s_v = se \tag{3.4}$$

where s_v is the vertical scale, s is the horizontal scale and e is the exaggeration. Once we have computed the size of our model we can build the model using any of the methods described in the following sections.

3.4.2   Sculpting a Malleable Model from Lidar Data

A simple way to create a malleable 3D model is to sculpt the landscape in polymer-enriched sand using projected contours derived from DEM as a guide. In this workflow we used a point cloud of elevation data in the LAS format (tile_0793_015_spm.las) and worked in the GRASS GIS location nc_spm_tl. Refer to the manual page of each relevant GRASS GIS command for more detailed explanation of the command syntax and parameters.

 To sculpt a malleable model from lidar data perform the following steps: Import the lidar points classified as ground (standard class 2) into GRASS GIS using the module *v.in.lidar*:

```
v.in.lidar -t input=tile_0793_015_spm.las output=ground \
    class_filter=2
```

Set the region to the spatial extent of the imported tile and resolution to 1 m. Then interpolate the DEM from the processed lidar data:

```
g.region vector=ground res=1
v.surf.rst -t input=ground elevation=dem tension=100 npmin=250 \
    dmin=2
```

Check the results, adjust the parameters, and rerun with the overwrite flag if necessary. Then compute the 1 m interval contours from the DEM:

```
r.contour input=dem output=contours_1m step=1
```

 We can now sculpt the terrain in polymer-enriched sand using the projected DEM and contours as a guide (Fig. 3.2). Once we have made at least a rough approximation of the form, we can use DEM differencing to critique and refine the modeled form (see Sect. 6.2.2).

3.4.3   CNC Routing a Topographic Model from Contour Data

Import the contours provided in .dxf file into Rhinoceros:

```
_Import contours.dxf _Enter
```

Zoom to the extent of all viewports:

```
Zoom All Extents
```

Select the contours and interpolate a terrain mesh:

```
_RtMeshTerrainCreate _Accept _Enter
```

Scale the model uniformly in the x, y, z direction to our chosen map scale by selecting the mesh, setting the origin to 0,0,0 and setting the scale factor:

```
_Scale 0,0,0 1/450 _Enter % 1/1500 ft
```

Optionally scale the model in the z-direction to exaggerate the relief by selecting the mesh, setting the origin to 0,0,0, setting the scale factor to 2, and setting the scale direction by drawing a line in the z-axis with ortho mode, using the gumball, or entering the coordinates 0,0,0 and 0,0,1:

```
_Scale1D 0,0,0 2 0,0,0 0,0,1 _Enter
```

Use RhinoCAM to generate the toolpath. Then CNC route the model.

3.4.4   CNC Routing Topographic and Surface Models from Lidar Data

Import the lidar data classified as ground (standard class 2) into GRASS GIS:

```
v.in.lidar -t input=tile_0793_015_spm.las output=ground \
    class_filter=2
```

Import the first return lidar data into GRASS GIS filtering unnecessary points:

```
v.in.lidar -t input=tile_0793_015_spm.las output=surface \
    return_filter=first class_filter=1,2,3,4,5,6,9
```

Set the region and resolution and interpolate the DEM from the ground points:

```
g.region vector=ground res=1
v.surf.rst -t input=ground elevation=dem tension=100 npmin=250 \
    dmin=2
```

Check the results, adjust the parameters, and rerun with the overwrite flag if necessary. With the region already set in the previous step, interpolate the DSM from the surface points:

```
v.surf.rst -t input=surface elevation=dsm tension=200 \
    smooth=0.5 npmin=120 dmin=2
```

Check the results, adjust the parameters, and rerun with the overwrite flag if necessary. Export the DEM to a space delimited text file with x, y, and z values:

```
r.out.xyz input=dem output=dem.xyz separator=space
```

Export the DSM to a space delimited text file with x, y, and z values:

```
r.out.xyz input=dsm output=dsm.xyz separator=space
```

Import the DEM text file into Rhinoceros as a point cloud. In the import options, set delimiter to 'space' and check 'create point cloud':

```
_Import dem.xyz
```

Import the DSM text file as a point cloud. In the import options, set delimiter to 'space' and check 'create point cloud':

```
_Import dsm.xyz
```

Zoom to the extent of all viewports:

```
Zoom All Extents
```

Select the DEM point cloud, interpolate a terrain mesh, and delete or hide the point cloud:

```
_RtMeshTerrainCreate _Accept _Enter
```

Select the DSM point cloud and interpolate a terrain mesh. Delete or hide the point cloud:

```
_RtMeshTerrainCreate _Accept _Enter
```

Scale both the DEM and DSM meshes uniformly in the x, y, z direction to our chosen map scale by selecting the meshes, setting the origin to 0,0,0 and setting the scale factor:

```
_Scale 0,0,0 1/1500 _Enter
```

Optionally scale both the DEM and DSM meshes in the z-direction to exaggerate the relief by selecting the meshes, setting the origin to 0,0,0, setting the scale factor to 2, and setting the scale direction by drawing a line in the z-axis with ortho mode, using the gumball, or entering the coordinates 0,0,0 and 0,0,1:

```
_Scale1D 0,0,0 2 0,0,0 0,0,1 _Enter
```

Optionally, create a vector curve delineating the border of the DEM or DSM:

```
_RtMeshFindBorder _Enter
```

Use RhinoCAM to generate the toolpaths. CNC route the models.

3.4.5   3D Printing Topographic and Surface Models from Lidar Data

Import the lidar data classified as ground (standard class 2) into GRASS GIS:

```
v.in.lidar -t input=tile_0793_015_spm.las output=ground \
    class_filter=2
```

Set the region to the given tile and set a relatively low resolution of 10 m. Interpolate the DEM from the ground points:

```
g.region vector=ground res=10
v.surf.rst -t input=ground elevation=dem tension=100 npmin=250
```

Export the DEM to a space delimited text file with x, y, z values:

```
r.out.xyz input=dem output=dem.xyz separator=space
```

Import the DEM text file into Rhinoceros as a point cloud. In the import options, set delimiter to 'space' and check 'create point cloud':

```
_Import dem.xyz _Enter
```

Zoom to the extent of all viewports:

```
Zoom All Extents _Enter
```

Select the DEM point cloud and interpolate a terrain mesh. Delete or hide the point cloud:

```
_RtMeshTerrainCreate _Accept _Enter
```

Scale the mesh uniformly in the x, y, z direction to our chosen map scale by selecting the mesh, setting the origin to 0,0,0 and setting the scale factor:

```
_Scale 0,0,0 1/3000 _Enter
```

Optionally scale the mesh in the z-direction to exaggerate the relief by selecting the mesh, setting the origin to 0,0,0, setting the scale factor to 2, and setting the scale direction by drawing a line in the z-axis with ortho mode, using the gumball, or entering the coordinates 0,0,0 and 0,0,1:

```
_Scale1D 0,0,0 2 0,0,0 0,0,1 _Enter
```

Create a vector curve delineating the border of the DEM:

```
_RtMeshFindBorder _Enter
```

Prepare the mesh for 3D printing. Select the border curve as the boundary and the DEM mesh as the mesh. Set offset to relative or absolute. Set a base height:

```
_RtMesh3dPrintPrepare
```

Export the selected mesh as a stereolithography file (.stl) and send it to the 3D printer:

```
_SaveAs
```

3.4.6 Casting a Malleable Topographic Model with a CNC Routed Mold Derived from Lidar Data

Import the lidar data classified as ground (standard class 2) into GRASS GIS:

```
v.in.lidar -t input=tile_0793_015_spm.las output=ground \
    class_filter=2
```

Set the region and resolution, then interpolate the DEM from the ground points:

```
g.region vector=ground res=1
v.surf.rst -t input=ground elevation=dem tension=100 npmin=250 \
    dmin=2
```

Check the results, adjust the parameters, and rerun with the overwrite flag if necessary. Export the DEM to a space delimited text file with x, y, z values:

```
r.out.xyz input=dem output=dem.xyz separator=space
```

Import the DEM text file into Rhinoceros as a point cloud. In the import options, set delimiter to 'space' and check 'create point cloud':

```
_Import dem.xyz _Enter
```

Zoom to the extent of all viewports:

```
Zoom All Extents _Enter
```

Select the DEM point cloud and interpolate a terrain mesh. Delete or hide the point cloud:

```
_RtMeshTerrainCreate _Accept _Enter
```

Optionally, create a vector curve delineating the border of the DEM:

```
_RtMeshFindBorder _Enter
```

Invert the DEM by rotating it 180 degrees on the z axis. Set the 'first reference point' by drawing a vertical line with ortho mode on and then set the 'second reference point' to '180' or by or entering the coordinates 0,0,0 and 0,0,1 and then 180 in the command line:

```
_Rotate 0,0,0 0,0,1 180 _Enter
```

Scale the mesh uniformly in the x, y, z direction to our chosen map scale by selecting the mesh, setting the origin to 0,0,0 and setting the scale factor:

```
_Scale 0,0,0 1/1500 _Enter
```

Optionally scale the mesh in the z-direction to exaggerate the relief by selecting the mesh, setting the origin to 0,0,0, setting the scale factor to 2, and setting the scale direction by drawing a line in the z-axis with ortho mode, using the gumball, or entering the coordinates 0,0,0 and 0,0,1:

```
_Scale1D 0,0,0 2 0,0,0 0,0,1 _Enter
```

Optionally, create a vector curve delineating the border of the DEM or DSM:

```
_RtMeshFindBorder _Enter
```

Use RhinoCAM to generate the toolpath, CNC route the inverted terrain model, and use this routed model as a mold to cast polymeric sand into a solid, malleable model of the topography.

References

Martisek, D., & Prochazkova, J. (2010). Relation between algebraic and geometric view on nurbs tensor product surfaces. *Applications of Mathematics, 55*(5), 419–430. Copyright - Institute of Mathematics of the Academy of Sciences of the Czech Republic, Praha, Czech Republic 2010; Last updated - 2014-08-22.

Modell, J., & Thuresson, S. (2009). Material composition and method for its manufacture. EP Patent App. EP20070794114.

Piegl, L., & Tiller, W. (1995). *The NURBS book.* New York: Springer.

Schodek, D., Bechthold, M., Griggs, K., Kao, K. M., & Steinberg, M. (2004). *Digital design and manufacturing.* Hoboken, New Jersey: John Wiley & Sons, Inc.

The GRASS GIS Community (2015). Contour lines to DEM [online]. Accessed 28.05.2015. http://grasswiki.osgeo.org/wiki/Contour_lines_to_DEM.

Chapter 4
Tangible Interactions

Geospatial models require various types of spatial data inputs, often with different attributes and geometries (i.e. continuous surfaces, points, lines, or polygons). To enable a broad range of applications, while keeping interactions tangible and intuitive, we use tangible objects such as wooden markers, wooden blocks, colored sand, and colored felt to specify various types of geospatial inputs. Depending upon the application, markers can be interpreted as viewpoints or waypoints, while cutout felt shapes of different colors can represent land cover or species habitat. Combining color and with changes in surface creates additional possibilities for tangibly interacting with geospatial models. We provide examples of different tangible interactions and explain the change detection, image segmentation, and image classification algorithms behind these methods.

4.1 Modes of Interaction

There are many ways to 3D sketch with Tangible Landscape (Fig. 4.1). Sculpting—either with bare hands or sculpting tools—is the most common way of interacting with an augmented clay or sand model. A single mode of interaction like this, however, limits the spectrum of geospatial modeling tasks. We can use tangible objects such as pins, wooden blocks, or pieces of textile such as felt to enable richer, more intuitive, and creative ways of interacting with geospatial models (Fig. 4.2).

In the following sections we describe various modes of interaction and explain how some of these objects can be detected and identified using simple approaches such as raster algebra, but also using more complex object recognition methods developed in the field of computer vision, which are implemented in GRASS GIS. Several examples of how tangible objects can be used in geospatial applications are included in this and the following chapters.

© The Author(s) 2018
A. Petrasova et al., *Tangible Modeling with Open Source GIS*,
https://doi.org/10.1007/978-3-319-89303-7_4

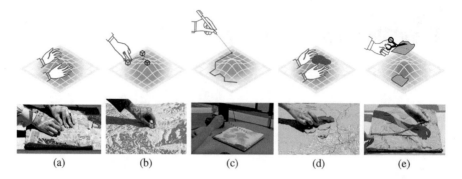

Fig. 4.1 Different modes of interaction used in Tangible Landscape: (**a**) sculpting the model with hand, (**b**) placing objects as markers, (**c**) drawing with laser pointer, (**d**) shaping and placing colored sand, and (**e**) cutting pieces of textile

Fig. 4.2 Examples of different tangible objects used for interaction: (**a**) pins for specifying point data, (**b**) wooden block for directions, and (**c**) pieces of felt for areal features

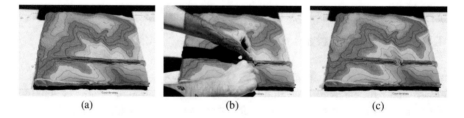

Fig. 4.3 3D sketching a levee breach with feedback on the resulting extent of flooding: (**a**) flooded landscape, (**b**) breaching the levee, and (**c**) updated flooding

4.2 3D Sculpting of Surfaces and Volumes

By sculpting a physical model made of clay or polymeric sand we can intuitively model topography (Fig. 4.3), which drives e.g., hydrological processes, solar radiation and visibility.

Physical models, however, can also represent more abstract surfaces, such as 3D raster cross-sections, cost, or probability surfaces. Figure 4.4 shows how we

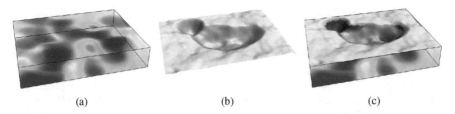

(a)   (b)   (c)

Fig. 4.4 Schema of interaction with 3D rasters: (**a**) 3D raster, (**b**) scanned surface of a physical model with excavated sand, and (**c**) cross-section of the scanned surface with 3D raster

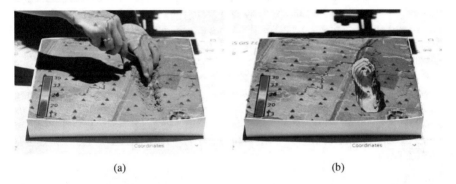

(a)   (b)

Fig. 4.5 Exploring 3D soil moisture dataset: (**a**) using hands and tools to show moisture values below ground and (**b**) projecting the cross-section of the scanned surface with soil moisture. Blue color represents highest and red lowest moisture percentage. The orthophoto, sampled sites, and flow accumulation are also projected onto the model as spatial context

can explore 3D raster data (also called volumes), which are typically difficult to visualize and interpret. 3D rasters in GRASS GIS represent trivariate continuous fields, such as soil or atmospheric properties, and are similar to 2D rasters, but with an additional z-dimension (called depth). One of the ways to visualize 3D rasters with Tangible Landscape is to create a physical model of the 3D raster as a box of sand and then modify the surface by excavating sand (Fig. 4.5). The surface is then used to compute the cross-section (2D raster) of the 3D raster (Fig. 4.4c) using the module *r3.cross.rast*:

```
r3.cross.rast input=map_3D elevation=scan output=cross_section
r.colors map=cross_section raster_3d=map_3D
```

An example of such interaction is shown in Fig. 4.5, where we visualize 3D soil moisture distribution measured on an agriculture field near Kinston, North Carolina (Duffera et al. 2007; Petrasova et al. 2014). Soil moisture and other soil properties were measured in different locations and depths, and processed into a 3D raster using trivariate interpolation implemented in *v.vol.rst* module. We can sculpt the sand (Fig. 4.5a) with our hands and then project the cross-section of the scanned surface with the 3D raster to visualize subsurface moisture levels, similar to on-site excavation and measurements (Fig. 4.5b). This method is an intuitive and natural way of exploring subsurface data and it represents an alternative to more abstract 3D computer visualization tools.

4.3   Detecting Markers

In Tangible Landscape applications point data can represent individual features such as trees or buildings, the position of an observer when modeling viewsheds, or the origin of fire when simulating wildfire spread. Alternatively, a group of points can be interpreted as a line or polygon, where the points represent vertices, allowing us to construct trails or least cost paths.

To specify these points on the landscape we can use small wooden blocks or pins, which we can stick in the sand model (Fig. 4.2a). The most reliable detection of these markers can be achieved by comparing the scan of the physical model before and after placing the marker, detecting the change in 3D, identifying markers based on their size, and then vectorizing the change raster into discrete vector points. Given the scanning resolution of the Kinect v2, the diameter and height of the marker should be at least 1.5 centimeters to enable reliable detection. It's best to avoid glossy and reflective materials which can hinder the scanning. The following Python functions can be used for marker detection during scanning. It computes the change raster and filters the change based on the vertical threshold and specified range of cells numbers. We use Python function `adjust_scan` previously defined in Sect. 6.2.2.

```python
def marker_detection(before, after, markers, h_thres, c_thres):
    adjust_scan(before, after, 'tmp_matched')
    gscript.mapcalc("{d} = if((({m}-{b})>{t1} && ({m}-{b})<{t2},
        1, null())".format(d='tmp_diff', m='tmp_matched',
        b=before, t1=h_thres[0], t2=h_thres[1]))
    gscript.run_command('r.clump', input='tmp_diff',
        output='tmp_diff_clump')
    stats = gscript.read_command('r.stats', flags='cn',
        input='tmp_diff_clump',
        sort='desc').strip().splitlines()
    # filter areas larger than specified number of cells
    cats = []
    for stat in stats:
        cat, val = stat.split()
        if float(val) < c_thres[1] and float(val) > c_thres[0]:
            cats.append(cat)
    if cats:
        rules = ['{c}:{c}:1'.format(c=cat) for cat in cats]
        gscript.write_command('r.recode',
            input='tmp_diff_clump', output=markers, rules='-',
            stdin='\n'.join(rules))
        gscript.run_command('r.volume', flags='f',
            input=markers, clump='tmp_diff_clump',
            centroids=markers)
```

Examples of different applications for markers are shown in Figs. 4.6, 4.7, 4.8. After the markers are detected, they can be used directly as input in many GRASS GIS modules, such as *r.drain* for computing a flow path (Fig. 4.6a), or they can be further processed to generate lines (Fig. 4.6b).

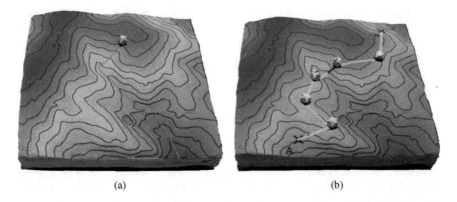

(a) (b)

Fig. 4.6 Examples of using markers to (**a**) explore where water drains from a point or (**b**) specify waypoints along a trail (with color representing slope along the trail as computed in Sect. 10.2.4)

(a) (b)

Fig. 4.7 Using magnetized markers on a CNC-routed model with magnetic primer to (**a**) draw boundaries of a polygon and to (**b**) rasterize an area

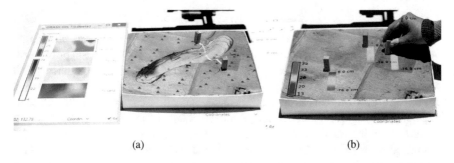

(a) (b)

Fig. 4.8 Examples of using markers to explore 3D soil moisture distribution: (**a**) visualizing vertical profiles of soil moisture and other soil properties with two markers and combining it with sand sculpting; (**b**) querying the 3D raster at any point with markers to obtain "soil core" information (Petrasova et al. 2014)

Alternatively, markers can be used to interact with 3D rasters as described in Sect. 4.2. We can use two markers to create a vertical profile (slice) of a 3D raster projected next to the model (Fig. 4.8a) or we can use multiple markers to query a 3D raster at the specified points and display the results beside them (Fig. 4.8b).

4.4   Detecting Color and Shape

When interacting with geospatial models we often need to input or modify areas to represent certain land cover, or more abstract concepts such as cost or probability. Using cloth or felt to define areas and their properties gives us a lot of flexibility—we can cut out various shapes from felt with different colors and move these pieces freely on the physical model. Based on the RGB information from the scanner (or any camera), the shapes and colors can be detected and interpreted to represent different phenomena.

The procedure consists of a calibration phase, when we assign different colors to different classes, followed by a classification phase, which determines the class of each RGB cell during scanning. We combine supervised classification with image segmentation to delineate clear outlines of each class.

For calibration we need to assign class numbers to different colors of felt, capture the physical model with felt as RGB image by the scanner, and then delineate these felt pieces on the scanned RGB maps as training areas. We can digitize the training areas and then convert them to raster map with the appropriate categories. Module *i.gensigset* then extracts spectral signatures from these training areas into a signature file, which then allows us to classify new RGB image with module *i.smap*. Module *i.group* is used here to collect raster map layers in an imagery group.

```
i.group input=color_r,color_g,color_b group=color subgroup=color
i.gensigset trainingmap=training group=color subgroup=color \
    signaturefile=signature
i.smap group=group subgroup=group signaturefile=signature \
    output=classified
```

Module *i.smap* classifies pixels into the specified classes and can provide an output raster map informing about the goodness of fit (Bouman and Shapiro 1994). Although the classification tries to avoid highly speckled results, the patches of felt are often not detected perfectly, resulting in misclassified edges or holes classified as different classes. This is where image segmentation can improve the process. Segmentation algorithms divide image into segments (also known as super-pixels) of uniform color values and delineate the borders between segments. We use the segmentation algorithm SLIC (Achanta et al. 2012) implemented in the GRASS GIS add-on *i.superpixels.slic*, which divides imagery group of raster maps into similarly sized segments with compact boundaries.

```
g.extension i.superpixels.slic
```

We independently compute the classification and segmentation and combine these results using zonal statistics in order to determine the most common class in each segment. In this way we safely filter out potentially misclassified pixels and still obtain continuous patches with more realistic compact boundaries.

```
def classify_colors(new, group, signature, compactness=2,
    threshold=0.3, minsize=10):
    # create segments
    gscript.run_command('i.superpixels.slic', input=group,
        output='tmp_segment', compactness=compactness,
        minsize=minsize)
    # classify scanned color maps in a group
    gscript.run_command('i.smap', group=group, subgroup=group,
        signaturefile=signature,
                        output='tmp_class', goodness='tmp_rej')
    # remove cells with high rejection value
    percentile = float(gscript.parse_command('r.univar',
        flags='ge', map='tmp_rejected')['percentile_90'])
    gscript.mapcalc('tmp_class_filtered = if(tmp_class < {p},
        tmp_class, null())'.format(p=percentile))
    # compute most common class in each segment
    gscript.run_command('r.mode', base='tmp_segment',
        cover=tmp_class_filtered, output=new)
```

In practice the calibration can be automated in Tangible Landscape plugin by creating the training areas raster and selecting it in the plugin's Analysis tab. Once we place pieces of felt on the model (Fig. 4.9a) we can click on the *Calibrate* button in Analysis tab to capture the RGB raster maps and extract the spectral signatures. The training layer is automatically hidden during the scanning to avoid interference. After calibration, the felt pieces of varied shapes can be detected and correctly classified using the above-mentioned procedure available in Tangible Landscape's library of functions.

The disadvantages of this method, which need to be considered for different applications, include the limited number of colors that can be reliably detected (around 6) and the decreased detection precision when multiple layers are projected on top of the physical model.

4.5 Combining Color and Elevation

The depth and color information from the scanner can be combined to create more abstract types of input data for geospatial models and simulations. Using colored moldable materials, such as colored sand or modeling clay, we can create colored volumes of varying shapes placed on top of a physical model (Fig. 4.10a). The color of the volume can represent a distinct category (e.g., landuse), while its height can be interpreted as a varying attribute of that category (e.g, elevation, density).

As an example, we use this technique to tangibly interact with a regional urban growth simulation by designing protected zones with limited development and

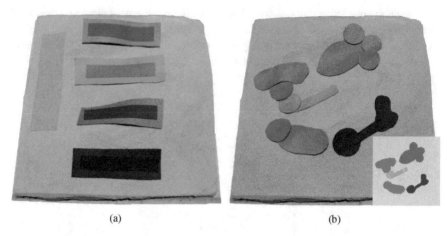

Fig. 4.9 Calibration and scanning of colored felt pieces: (**a**) placing felt pieces on projected training areas; (**b**) detecting the color and shape of felt pieces. The resulting raster is shown in the inset in the bottom right corner.

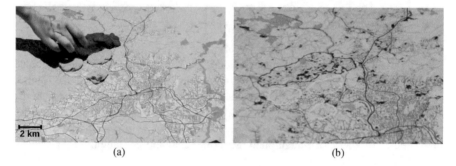

Fig. 4.10 Using colored sand to design new urban development zones in Asheville: (**a**) red increases probability of new development and green represents protected zones; (**b**) sand is detected and classified into zones and resulting computational scenario shows dark red patches of new urban development with high concentration in the area with red sand (zoomed in figure)

desired development zones. We use FUTURES, a stochastic, patch-based urban growth model implemented as the set of GRASS GIS add-on modules *r.futures* (Meentemeyer et al. 2013; Petrasova et al. 2016), which simulates land conversions in discrete patches based on a probability surface.

By creating red zones (which support new development) and green zones (which protect existing land cover) using colored sand (Fig. 4.10a), we can modify the probability surface and create new scenarios for future development. By changing the height of each zone we can locally increase or decrease the probability of development. Figure 4.10b shows one of the scenarios for anticipated urban growth in next 20 years. The green and red outlines represent the designed zones, and the dark red patches concentrated in the red zone represent simulated development.

The following code snippet shows a particular way to implement this approach. We first identify the designed zones based on their elevation difference from the physical model beyond a threshold value. There are multiple ways to identify the category of each zone based on its color. For simplicity's sake we demonstrate a fast method that distinguishes two distinct colors. Alternatively, multiple colors can be detected using the procedure described in Sect. 4.4.

For the purpose of *r.futures* module inputs, in this code snippet the elevation difference of all zones is scaled from 0 to 1 based on the maximum elevation (using the *graph* function for linear interpolation implemented in *r.mapcalc*). However, other applications of this interaction method may require different scaling. The identified zones can be then passed into the *r.futures* model to run the urban growth simulation.

```
def colored_sand(before, after, color, thres):
    adjust_scan(before, after, 'tmp_matched')
    gscript.mapcalc('change = if((({m}-{b})>{t}, {m}-{b},
        null())'.format(m='tmp_matched', b=before, t=thres))
    gscript.mapcalc('change_bin = if(change, 1, null())')
    gscript.run_command('r.clump', input='change_bin',
        output='clump')
    gscript.run_command('r.stats.zonal', base='clump',
        cover=color+'_r', output='red_mean', method='average')
    gscript.run_command('r.stats.zonal', base='clump',
        cover=color+'_g', output='green_mean', method='average')
    # identify red zones, convert to vector
    gscript.mapcalc('red = if(red_mean > green_mean, 1,
        null())')
    gscript.run_command('r.to.vect', input='red', output='red',
        type='area', flags='s')
    # scale magnitude of elevation change to 0 to 1
    max_change = gscript.raster_info('change')['max']
    gscript.mapcalc('red_magnitude = if(red_mean > green_mean,
        graph(change, {}, 0, {}, 1), null())'.format(thres,
        max_change))
    # similarly for green zones
```

4.6   Direction Marker

Directionality is a common property of many terrain visualization techniques, solar radiation models, or dynamic simulations of widlfire or disease spread. In Tangible Landscape applications we can input direction using a dedicated tangible object (Fig. 4.2b), which can be implemented as a wooden block of approximately 1.5 × 1.5 × 5 centimeters preferably with a needle in the center to hold the object on the physical model. One half of the marker is painted with darker color specifying the main direction in the same way a compass needle represents the north direction. To detect this object we combine the depth and color information from the scanner—

the elevation difference (see Sect. 4.3) identifies the position of the direction marker and the color difference in the marker surface determines the direction. Instead of the full classification procedure introduced in Sect. 4.4 we apply a simplified method to classify each pixel of the marker as either of the two classes based on the brightness threshold. This simple method typically works even when we project other data, while using the marker. The choice of brightness threshold, however, depends on the projected data. By computing a centroid of each class using *r.volume* module, we obtain an approximate direction vector:

```
def direction_marker(before, after, group, marker, h_thres,
    b_thres):
    adjust_scan(before, after, 'tmp_matched')
    gscript.mapcalc('{r} = if((({m}-{b})>{t}, 1,
        null())'.format(m='tmp_matched', b=before, t=h_thres,
        r='change'))
    gscript.mapcalc('{r} = if({c} && ({g}_r + {g}_g + {g}_b) /
        3. >= {t}, 1, 2)'.format(r=marker, c='change',
        t=b_thres, g=group))
    gscript.run_command('r.volume', flags='f', input=marker,
        clump=marker, centroids=marker)
```

Once we have computed the two centroids, we can use this information for various tasks, such as computing the direction and converting it to degrees in a specific convention, for example by mathematical conventions degrees are measured from the east in counter-clockwise direction, while many other applications measure angles clockwise from north. From the centroids we can also obtain the approximate center of the direction marker.

```
r = gscript.read_command('v.out.ascii',
    input=marker).strip()
p = []
for point in r.splitlines():
    x, y, c = point.split('|')
    p.append((float(x), float(y)))
# needs from math import atan2, pi
angle = atan2(p[1][1] - p[0][1], p[1][0] - p[0][0])
angle_deg = angle * 180 / pi
center = (p[0][0] + p[1][0]) / 2., (p[0][1] + p[1][1]) / 2.
```

Figure 4.11 demonstrates how the direction marker dynamically changes the azimuth of shaded relief. In this example the particular position of the marker is not important, whereas in viewshed modeling as shown in Fig. 4.12, the position specifies the observer's location and the direction represents the orientation of the observer. The 360° viewshed map is masked to project the human field of view.

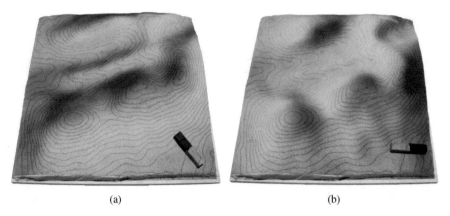

(a)   (b)

Fig. 4.11 Using the direction of the marker to change azimuth of the sun for shaded relief computation: (**a**) 320° and (**b**) 90° clockwise from north

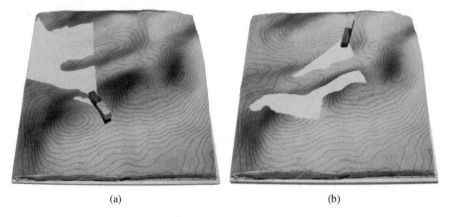

(a)   (b)

Fig. 4.12 Using the position and direction of the marker to compute visibility with the field of view constrained by the given angle

References

Achanta, R., Shaji, A., Smith, K., Lucchi, A., Fua, P., & Süsstrunk, S. (2012). SLIC superpixels compared to state-of-the-art superpixel methods. *IEEE Transactions on Pattern Analysis and Machine Intelligence, 34*(11), 274–2282.

Bouman, C. A., & Shapiro, M. (1994). A multiscale random field model for Bayesian image segmentation. *IEEE Transactions on Image Processing, 3*(2), 162–177.

Duffera, M., White, J. G., & Weisz, R. (2007). Spatial variability of Southeastern U.S. Coastal Plain soil physical properties: Implications for site-specific management. *Geoderma, 137*(3–4), 327–339.

Meentemeyer, R. K., Tang, W., Dorning, M. A., Vogler, J. B., Cunniffe, N. J., & Shoemaker, D. A. (2013). FUTURES: Multilevel simulations of emerging urban-rural landscape structure using a stochastic patch-growing algorithm. *Annals of the Association of American Geographers, 103*(4), 785–807.

Petrasova, A., Harmon, B., Mitasova, H., & White, J. (2014). Tangible exploration of subsurface data. Poster presented at 2014 Fall Meeting, AGU, San Francisco, CA, 15–19 December.

Petrasova, A., Petras, V., Van Berkel, D., Harmon, B. A., Mitasova, H., & Meentemeyer, R. K. (2016). Open source approach to urban growth simulation. *ISPRS - International Archives of the Photogrammetry, Remote Sensing and Spatial Information Sciences, XLI-B7*(July), 953–959.

Chapter 5
Real-Time 3D Rendering and Immersion

People's perception and experience of landscape plays a critical role in the social construction of these spaces—in how individuals and societies understand, value, and use landscapes. Perception and experience should, therefore, be an integral part of environmental modeling and geodesign (Steinitz 2012; Nassauer 1997; Gobster et al. 2007). With the natural interaction afforded by Tangible Landscape and the realistic representations afforded by Immersive Virtual Environments (IVEs) experts and non-experts can collaboratively model landscapes and explore the environmental and experiential impacts of "what if" scenarios (Smith 2015; Tabrizian et al. 2018). We have paired GRASS GIS with Blender, a state-of-the-art 3D modeling and rendering program, to allow real-time 3D rendering and immersion. As users manipulate a tangible model with topography and objects, geospatial analyses and simulations are projected onto the tangible model and perspective views are realistically rendered on monitors and head-mounted displays (HMDs) in near real-time. Users can visualize in near real-time the changes they are making with either bird's-eye views or perspective views from human vantage points. While geospatial data is typically visualized as maps, axonometric views, or bird's-eye views, human-scale perspective views help us to understand how people would experience and perceive spaces within the landscape.

5.1 Blender

Blender is a free and open source software for 3D modeling, rendering and game design (Blender Online Community 2016). We use this software to 3D model and 3D render geospatial data in near real-time. It has an easy-to-use Python API for automating procedural 3D modeling workflows. It supports realtime viewport

© The Author(s) 2018
A. Petrasova et al., *Tangible Modeling with Open Source GIS*,
https://doi.org/10.1007/978-3-319-89303-7_5

shading[1] with a sufficient degree of realism. Blender has a GIS add-on for importing and processing georeferenced data in raster and vector formats. There are also several plugins for displaying the viewport content in head-mounted displays (HMDs) for realtime immersive interaction with 3D models.

5.2   Hardware and Software Requirements

With three additional components Tangible Landscape setup can support 3D rendering and immersive display: a computer with network connection, a monitor, and a head-mounted display (Fig. 5.1). For optimal performance we recommend a Virtual

Fig. 5.1 Tangible Landscape setup with 3D modeling and rendering components including Blender, a computer monitor, and an Oculus Rift headset

---

[1] Viewport shading refers to drawing 3D geometries and computing their shading (e.g., textures and reflection) and lighting (e.g., cast and received light and shadow).

Reality (VR) ready computer with the most recent version of Blender. Blender and GRASS GIS can be run on the same computer if that the computer has enough computing power and its graphics card supports at least three display outputs. Display outputs are needed for Tangible Landscape's operator's display, Tangible Landscape's projection, and Blender's rendering display. Although Blender is multiplatform software, the choice of operating systems may be limited when the immersion is required. We use the *Virtual Reality Viewport* add-on (Felinto 2015), currently available for MS Windows, which supports the Oculus Rift DK2 and Oculus runtime (0.8). There are, however, other VR add-ons for Blender (e.g., OpenHMD) that can be compiled with Blender to enable multiplatform HMD support (Open HMD Team 2016).

5.3   Software Architecture

Blender and GRASS GIS are loosely coupled through file-based communication established via a local wireless or Ethernet connection. GRASS GIS exports the spatial data as a standard raster, a vector, or a text file containing coordinates into a specified directory typically called *Watch* (Fig. 5.2). The Tangible Landscape Blender plugin (`modeling3D.py`)—implemented and executed inside Blender— constantly monitors the directory for incoming information. Examples of spatial data include a terrain surface (raster), water bodies (3D polygons or rasters), forest patches (3D polygons), a camera location (3D polyline, text file), and routes (3D polylines). Upon receiving this information, the file is imported using the BlenderGIS add-on. Then the relevant modeling and shading procedures for updating an existing 3D object or creating a new 3D object are applied. The

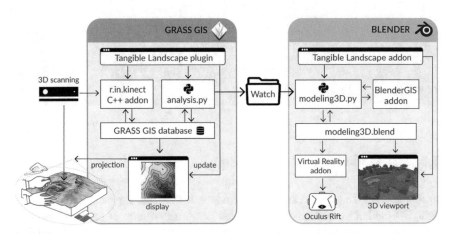

Fig. 5.2 Tangible Landscape's software architecture couples tangible interaction, 3D scanning, geospatial computation in GRASS GIS with 3D modeling and rendering in Blender

adaptation procedure applied depends upon the type of spatial data and is handled by a module called *adapt*. All 3D elements in the scene (i.e. objects, lights, materials, and cameras) reside in a Blender file (modeling3D.blend).

5.4   File Monitoring

File monitoring is handled through a native function of Blender called the *Modal Timer Operator*. We use this particular module instead of Python libraries for monitoring such as *Watchdog*, because these libraries can interfere with Blender's ability to run multiple operators at once and update different parts of the interface.[2] The following snippet demonstrates the structure of the modal timer function. In this example, the monitored folder is inventoried every second and when a terrain surface is the detected (e.g., terrain.tif), the *adapt* module executes to update the terrain model.

```
def modal(self, context, event):
    if event.type in {"RIGHTMOUSE", "ESC"}:
        return {"CANCELLED"}

    if event.type == "TIMER":
        if self._timer.time_duration != self._timer_count:
            self._timer_count = self._timer.time_duration
            fileList = (os.listdir(watchFolder))

            if terrainFile in fileList:
                adapt().terrain()

        # execute the timer for the first time
        def execute(self, context):
            wm = context.window_manager
            wm.modal_handler_add(self)
            self._timer = wm.event_timer_add(1, context.window)
            return {"RUNNING_MODAL"}

        def cancel(self, context):
            wm = context.window_manager
            wm.event_timer_remove(self._timer)
```

5.5   3D Modeling and Rendering

In this section we provide an overview of the techniques for handling geospatial data, 3D modeling, and rendering in Blender. A description of each technique is followed by a Python code snippet that can be used to program procedural modeling

---

[2] See https://docs.blender.org/api/blender_python_api_2_62_2/info_gotcha.html.

and shading workflows for various types of geospatial data (see Sect. 5.6 for sample workflows). All the code snippets use Blender's Python API (*bpy*) and assume that it has been imported.

```
import bpy
```

5.5.1   Handling Geospatial Data

Importing Geospatial Data in Blender We use Blender GIS add-on *importgis* functions to import raster and vector formats. The coordinate reference system (CRS) of the spatial data should be specified (using an EPSG code) in both the add-on configuration and the import function. A raster can be imported as an interpolated 3D surface (option *DEM*), a point cloud (option *Raw DEM*), or a texture to be draped onto an existing mesh (option *On mesh*). The following snippet provides an example for importing a digital elevation model (DEM) as a 3D mesh using the *DEM* method.

```
import os, bpy
inputFile = os.path.join(dirPath, 'terrain.tif')
bpy.ops.importgis.georaster(filepath=inputFile,
    importMode="DEM", subdivision="subsurf",
    rastCRS="EPSG:3358")
```

Vector formats are used for points and linear features that can be linked to interactions with tangible markers, to lines drawn by laser pointer, or to simulations such as routes and trails. BlenderGIS can not only import geometry, but also vector attributes such as elevation and height, which used for extrusion. If exported vector features contain z values (3D polyline or 3D polygons), it is not necessary to specify the elevation parameter. It should be also noted that, unlike the raster DEMs, the imported shape features are by default lack surfaces and cannot be rendered unless an extrusion parameter is specified. If a shapefile contains multiple features (e.g., buildings or points), the *seperateObject* parameter can be used to break the vector layer into discrete 3D objects. The following snippet imports a shapefile containing multiple points with height and elevation attributes:

```
inputFile = os.path.join(dirPath, 'points.shp')
bpy.ops.importgis.shapefile(filepath=inputFile,
    fieldElevName="height", fieldObjName='Name',
    separateObjects=True, shpCRS='EPSG:3358')
```

Spatial features with closed boundaries such as water bodies and vegetated patches can be exported as either vector features (polygons) or rasters. There is, however, is a trade-off that should be considered when choosing which format to use. Shapefiles produce more accurate edges and can be imported as discrete objects, but fit less accurately on rough or undulating topography. Furthermore, it is difficult to rapidly and continuously transmit shapefiles because they are composed of multiple files (i.e. .shp, .prj, .dbf, and .shx). Rasters on the other hand are easier to transfer, import, and fit accurately on topography.

Exporting Geospatial Data from GRASS GIS   Raster features can be exported using the module *r.out.gdal*. When exporting digital elevations models, specify the GeoTIFF format with the data-type set to 32bit float:

```
r.out.gdal -cf input=dem out=path/output.tif type=Float32 \
   format=GTiff create="TFW=YES"
```

Maps with RGB information such as orthophotographs or simulation outputs (e.g., waterflow) should be exported as PNG or JPG formats:

```
# decompose raster to red, green and blue channels
r.rgb input=texture red=red green=green blue=blue
i.group group=texture_group input=red,green,blue
r.out.gdal input=texture_group output=path/output.PNG \
   format=PNG type=Byte createopt="WORLDFILE=YES"
```

Vector data including points, lines, and boundary features can be exported using the module *v.out.ogr*:

```
v.out.ogr input=vector output=path/output.shp \
   format="ESRI_Shapefile" lco="SHPT=POINT"
```

Depending on the type of data, the `lco` option needs to be adjusted to `SHPT=ARC` when exporting lines or to `SHPT=POLYGON` when exporting areas. Furthermore, when vector data has Z coordinates, the additional letter `Z` needs to be appended with `SHPT=POINTZ`.

5.5.2   Object Handling and Modifiers

Object Management   While in GIS functions are typically applied by specifying a layer as an input, in Blender modifications such as moving, deleting, hiding, and rotating are applied to any object that is *selected*. This makes object management an integral task for real-time modeling given that the data is continuously transmitted between the GRASS GIS and Blender, and multiple spatial data may be processed simultaneously. In Blender the existing data (i.e., materials, objects, meshes, textures, lights, and cameras) and their status can be retrieved with the *bpy.data* module and operations can be applied using the *bpy.ops* module.

The following snippet demonstrates the procedure for checking if a terrain object resides in the scene, deselecting any previously selected objects, selecting the terrain object, and removing it from the scene before importing and processing the new terrain object:

```
if bpy.data.objects.get("terrain"):
    bpy.ops.object.select_all(action='DESELECT')
    bpy.data.objects["terrain"].select = True
    bpy.ops.object.delete()
```

Object Transformation   Because the 3D scene in Blender and GRASS GIS use the same georeferencing system, it is possible to communicate the changes in position of tangible objects and their corresponding 3D objects using x, y, z coordinates exported by GRASS GIS. The following snippet relocates a 3D model of a building when a user moves a wooden marker on the tangible model. Object coordinates are transferred as text files.

```
buildingObj = bpy.data.objects["Building"]
buildingObj.location = [X, Y, Z]
```

The snippet below demonstrate how to retrieve the start and endpoint coordinates of an imported polyline. The line object is converted to a mesh and the vertices' locations are retrieved.

```
lineObj= bpy.data.objects["line"]
mesh = lineObj.to_mesh(bpy.context.scene, apply_modifiers=True,
    settings='PREVIEW')
startCoord = mesh.vertices[0].co
endCoord = mesh.vertices[1].co
```

Shrink Wrapping   Shrink wrapping is a modifier in Blender that *wraps* an object onto the surface of another object. The modifier moves each vertex of the selected object to the closest position on the surface of the given mesh. We apply this technique for draping a two-dimensional surface on the terrain because it properly aligns edges and avoids floating or drowning objects. The following snippet provides an example of using shrink wrap modifier to drape an imported 2D patch onto the terrain object (i.e. target object).

```
plane = bpy.data.objects["plane"]
plane.select = True
bpy.data.objects["terrain"].select = True
bpy.ops.object.modifier_add(type='SHRINKWRAP')
plane.modifiers['Shrinkwrap'].target =
    bpy.data.objects["terrain"]
plane.modifiers["Shrinkwrap"].wrap_method = "NEAREST_VERTEX"
plane.modifiers["Shrinkwrap"].use_keep_above_surface = True
```

Particle Systems   Often tangible geospatial modeling we deal with spatial features that are composed of a large number of very small individual objects. Examples include populating city blocks with buildings or patches of forest with specific plants. In a tangible model the boundaries of such features can be demarcated using felt pieces, colored sand, or lines drawn by laser pointer. In a 3D model they can be represented as a a point cloud of particles arranged either randomly or by rules. In Blender these collections of particles are generated using the *particle system* modifier. The *particle system*'s parameters include the count of the particles, particle size and rotation, randomness in distribution, size and rotation, random use of a group of objects, and physics that define the relationships between particles such as deflection. The following snippet demonstrates the random distribution of maple trees in a patch. Three maple trees of different ages (young, middle aged, and adult) are randomly drawn from a group object and populated with a random size, rotation and distribution within a boundary object. By calculating the area of an object in Blender, the population density can be adjusted as function of the count per area unit (in this case one tree every $200\,m^2$).

```
patchobj = bpy.data.objects ["patch"]
groupobj = bpy.data.groups["Maple trees"]

bpy.ops.object.particle_system_add()
pset1 = obj.particle_systems[-1].settings
pset1.name = 'TreePatch'
pset1.type = 'HAIR'

pset1.render_type = 'GROUP'
pset1.dupli_group = groupobj
pset1.use_group_pick_random = True

pset1.use_emit_random = True
pset1.lifetime_random = 0.0
pset1.emit_from = 'FACE'
pset1.count = getArea(patchobj) / 200
pset1.use_render_emitter = True

pset1.use_emit_random = True
pset1.userjit = 70
pset1.use_modifier_stack = True
pset1.hair_length = 0.6

pset1.use_rotations = True
pset1.rotation_factor_random = 0.02
pset1.particle_size = 1
pset1.size_random = 0.4
```

5.5.3 3D Rendering

3D rendering is the automatic generation of images from 3D models. The software for rendering—the render engine—controls the materials and lighting, how the objects are drawn, shaded and lit in the viewport (viewport shading), and the realism and quality of renderings.[3] While viewport shading is real-time, a single high quality production render may take hours or even days to compute.

Blender has two built-in render engines—Blender Render and Cycles—and also supports external render engines including LuxRender, Maxwell, V-Ray, and Octane. Blender Render is a scanline rasterization engine for non photo-realistic rendering, while Cycles is a physically based, path-tracing engine for photoreal-

---

[3]https://docs.blender.org/manual/en/dev/render/introduction.html.

(a)

(b)

(c)

Fig. 5.3 Three modes of 3D rendering: (**a**) viewport display with Blender Render engine, (**b**) viewport display with Cycles engine, and (**c**) full render with Cycles engine

istic rendering with global illumination. While Blender Render does not support raytraced lightning and caustics, its speed is useful for real-time viewport shading. Blender 2.8 will include the new real-time, physically based render engine Eevee. We recommend using the Blender Render engine throughout the modeling process because of its speed and then Cycles for rendering the final production graphics because of its quality (Fig. 5.3). The following snippets demonstrate the commands for switching between the active render engine:

```
bpy.context.scene.render.engine = "CYCLES"
bpy.context.scene.render.engine = "Blender"
```

5.5.4 Materials

Objects' materials directly influence the appearance and realism of a 3D scene. Materials play in important role in generating lifelike representations from abstract map features and tangible objects. We briefly discuss two basic components of modeling materials, namely *Shading* and *Texture mapping*. Shading (or coloring) is a technique for adjusting the base color (as modified by diffusion and specular reflection) and light intensity of an object's surface. Texture mapping is the process of draping images and patterns to add detail to the surfaces. Examples include draping an aerial image (i.e., orthophotograph) or a grassy texture onto the terrain or assigning a rippling wave texture to a water surface.

With the *Cycles* render engine shading and texture parameters are stored in network of *Nodes*, which define the surface and volumetric properties of the material. A water material, for example, can be defined by surface property nodes such as transparency and glossiness and volumetric property nodes such as ripple effects and wave textures (Fig. 5.4). While it is possible to generate an entire material using Python code, this can be processing and time intensive since the data related to the material network (nodes) and attributes are stored in the `bpy.data` object. Therefore, we recommend assigning or modifying previously prepared materials to reduce the processing time for realtime modeling. The following snippets demonstrates the procedure for generating a simple terrain material, assigning the material to a terrain object, and replacing an image in the texture node of an existing material (Fig. 5.5a):

```
filePath = os.path.dirname(bpy.path.abspath("//"))
orthoFile = os.path.join(filePath, 'ortho.png')
matName = "orthoMat"
mat = bpy.data.materials.new(matName)
obj.data.materials.append(mat)
# Get material tree, nodes and links
mat.use_nodes = True
node_tree = mat.node_tree
nodes = mat.node_tree.nodes
links = node_tree.links
for node in nodes:
    nodes.remove(node)
# Create a diffuse node, a texture node, and an output node
diffuseNode = node_tree.nodes.new("ShaderNodeBsdfDiffuse")
orthoNode = node_tree.nodes.new("ShaderNodeTexImage")
orthoNode.image = bpy.data.images.load(orthoFile)
outputNode = node_tree.nodes.new("ShaderNodeOutputMaterial")
# Create the links
links.new(orthoNode.outputs["Color"],
    diffuseNode.inputs["Color"])
links.new(diffuseNode.outputs["BSDF"],
    outputNode.inputs["Surface"])
```

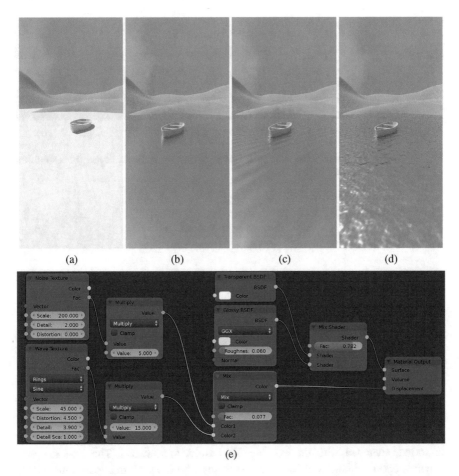

(a)   (b)   (c)   (d)

(e)

Fig. 5.4 Renderings of a water material: (**a**) with color, (**b**) with transparency and glossiness, (**c**) with a wave texture, and (**d**) with a noise (ripple) texture. The water material was created using the nodes shown in the (**e**) Node Editor. Noise and Wave texture nodes were assigned to the surface volume to create ripple and wave effects. Math multiply nodes were used to adjust the magnitude of the effects. Transparency and Glossy shaders were assigned to the surface

The following script assigns an existing material (orthoMat) to a terrain object (Fig. 5.5a).

```
obj = bpy.data.objects["terrain"]
mat = bpy.data.materials.get("orthoMat")
obj.data.materials.append(mat)
```

The snippet below replaces the existing surface texture of the terrain model (an aerial image) with a viewshed map computed in GRASS GIS (Fig. 5.5b):

```
filePath = os.path.dirname(bpy.path.abspath("//"))
viewshedFile = os.path.join(filePath, 'viewshed.png')
```

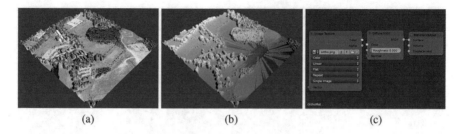

(a)   (b)   (c)

Fig. 5.5 Assigning surface textures using a diffuse shader: (**a**) an aerial image, (**b**) a viewshed map, and (**c**) the Material tree in the node editor

```
mat = (bpy.data.materials.get("OrthoMat"))
texNode = mat.node_tree.nodes["Image Texture"]
texNode.image = bpy.data.images.load(viewshedFile)
```

5.6 Workflows

In this section we describe workflows for importing, 3D modeling and shading spatial features that are created or modified through tangible modeling. The automated procedure for adapting each spatial feature in Blender is implemented as a function in the *adapt* module.

Terrain Tangible manipulations of the physical terrain model can be communicated from GRASS GIS to Blender with a digital elevation model that is iteratively imported, swapped with the existing 3D terrain, and shaded. The import speed depends upon resolution of the raster. In cases where terrain manipulation follows other tasks that involve adding objects (e.g., planting trees), then an additional shrink wrapping step should be applied to drape all the above-surface objects back onto the new terrain (see Sect. 5.5.2).

1. Check bpy.data to determine if the terrain object already exist in the scene. If it does, then delete it.
2. Import the new terrain raster (Fig. 5.6a).
3. Convert the imported feature to a Mesh object. This conversion enables further modifications of terrain in the subsequent steps.
4. Add side fringes to the terrain object. Fringes enhance appearance of the terrain in the bird's-eye view mode (Fig. 5.6b).
5. Assign the "Grass" material to the terrain and the "Dirt" material to the fringes (Fig. 5.6c).

(a)   (b)   (c)

Fig. 5.6 The process for modeling a terrain feature: (**a**) after importing the GeoTIFF raster, (**b**) after adding fringes, and (**c**) after assigning the grass material

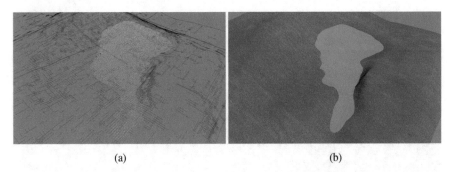

(a)   (b)

Fig. 5.7 Modeling a water feature: (**a**) the imported GeoTIFF raster, and (**b**) the mesh after assigning the "Water" and "Grass" materials

Water Features When a user sculpts the tangible model, a depression filling algorithm (*r.fill.dir*) in GRASS GIS can simulate water features such as lakes and ponds. These features can be exported either as a 3D polygon shape or a GeoTIFF. We recommend using a relatively high-resolution raster to minimize raised edges or gaps between the outer boundary of the imported feature and the basin, especially with rough terrain.

1. Check if the water object already exist. If it does, then delete it.
2. Import the water raster (Fig. 5.7a).
3. Assign the "Water" material to the water object (Fig. 5.7b).

Forest Patches Users can tangibly model patches of trees using felt pieces (See Fig. 4.9) or delineate a single species with a colored wooden marker. Tree patches are scanned and classified in GRASS GIS and exported to Blender as 3D polygons. In Blender a particle system modifier is used to populate predefined tree models in the imported patches according to predefined distribution rules. Vegetation models can be obtained from 3D model libraries such as Xfrog[4] or can be modeled using procedural plant generation software or add-ons such as The Grove.[5]

---

[4]http://xfrog.com/.

[5]https://www.thegrove3d.com/.

Fig. 5.8 Importing polygon features and populating four types of trees based on patch classification: (**a**) 3D models of individual trees, (**b**) a wireframe representation of patches after importing, and (**c**) patches with the particle system modifier applied

1. Import the patch shapefile (Fig. 5.8b).
2. Calculate the patch area.
3. Check the type of patch (based on the file name) and assign the particle system modifier to populate the designated tree using distribution rules for the density and distribution of objects (Fig. 5.8c).
4. Assign a transparent material to the patches.

Trail Routes and trails can be tangibly modeled using markers (Fig. 4.6b) and exported as 3D polylines. In Blender the predefined profile of the pathway is extruded along the imported features using the bevel modifier[6] (Fig. 5.9a). To better extrude sharp bends and curves, we recommend smoothing the 3D polyline feature in GRASS GIS using *v.generalize* command before exporting.

---

[6]https://docs.blender.org/manual/en/dev/modeling/modifiers/generate/bevel.html.

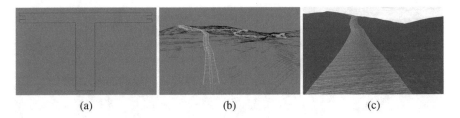

(a)   (b)   (c)

Fig. 5.9 Modeling a trail feature: (**a**) the imported polyline feature displayed as a curve object in the wireframe display, (**b**) the trail after applying the Bevel modifier using a T profile curve, and (**c**) the boardwalk after assigning the wood plank material

1. Check if the trail already exist. If it does, then delete it.
2. Import the trail shapefile (3D polyline).
3. Convert the imported object to Curve. The conversion enables the bevel modifier to be applied in the next step.
4. Extrude the T profile along the curve.
5. Assign the wood planks material to the extruded feature.

Camera Users can tangibly explore human views by placing a colored wooden marker on the tangible model (Fig. 4.2b). In GRASS GIS the view marker is exported as a polyline feature. In Blender the first vertex is interpreted as the camera location and the second point represents the view target.

1. Import the 3D polyline.
2. Retrieve the coordinates of the line's first and second vertices.
3. Move the camera and target to the retrieved coordinates.
4. Move the camera's Z coordinate to eye level (1.65 m).

5.7   Realism and Immersion

5.7.1   Realism

The level of realism is an important aspect of visualization. Both abstract and photorealistic representations can be equally useful depending on the purpose of visualization, the audience, and computational resources available. Abstract visualizations—features with less geometric and texture complexity—are less cognitively and computationally demanding making them useful for rapid prototyping and the early, conceptual phases of the design process. Abstract visualization can also appeal to younger age groups and are thus useful for education. Photorealistic representations, on the other hand, are very useful for representing the experience and aesthetics of a design.

(a)   (b)

Fig. 5.10 Viewport rendering of the Blender scene with two modes of realism: (**a**) low-poly mode and (**b**) realistic mode

We implemented a function that enables users to select between either a realistic or abstract (low-poly) mode of visualization at anytime during the tangible modeling process (Fig. 5.10). This was done by creating alternate worlds (sky and background elements), objects (e.g., low-poly trees), and textures. In other words every component of the 3D scene including lights, objects and background have a low-poly and a realistic instance. The following snippets include a function that loops through the elements in the scene and swaps them with their alternate object. Swapping the world based on the mode of realism defined by the user:

```
def updateWorld(mode):
    newWorld = mode
    bpy.context.scene.world = bpy.data.worlds[newWorld]
    self.world= bpy.data.worlds[newWorld]
```

Swapping the realistic and abstract trees in an instance of a particle system assigned to the forested patches based on the mode of realism defined by the user:

```
def changeRealism(mode):
    for obj in bpy.data.objects:
        if "patch_" in obj.name and obj.particle_systems:
            newParticle = mode + "_" + obj.name.split("_")[1]
            setting = obj.particle_systems[0].settings
            setting.dupli_group = bpy.data.groups[newParticle]
```

5.7.2   Virtual Reality Output

User can access the VR output using the add-on panel located on the Blender's 3D viewport's tool shelf. The add-on converts the scene camera to panoramic display and broadcast it as a stereoscopic image onto the head-mounted display. It is possible to display the scene in the HMD while the Tangible Landscape plugin is in *watch mode*. However, this can slow down the system and occasionally causes crashes for more complex scenes especially in realistic mode. For a better VR experience, we recommend using one feature at a time, i.e., stopping the watch mode before displaying VR and vice versa.

Fig. 5.11 The Tangible
Landscape add-on in Blender

5.8 Tangible Landscape Add-on in Blender

In this section we describe the main features and components of the Tangible
Landscape GUI in Blender. After installation the add-on panel can be accessed from
the 3D view's tool shelf (Fig. 5.11). The *Watch Mode* button activates the modal
timer function and waits for Tangible Landscape to copy files to the *Watch* folder.
The second panel *Camera options* allows user to toggle between the following
four camera types: a human view camera linked to the tangible view marker (See
Sect. 5.6), preset bird's-eye views, preset human views, and a bird's-eye view for
an animated orbiting camera. The preset bird's-eye and human cameras are linked
to the 3D viewport allowing users to navigate the 3D scene (by mouse) and adjust
and revisit their preferred views. The *Rendering and realism* panel includes buttons
for selecting between the Cycle and Blender Render engines and between low-poly
and realistic representations. Thanks to Blender's flexible and accessible interface
design, users can modify existing features or add new ones to accommodate specific
project needs. Some examples include enabling atmospheric effects (e.g., mist, rain,
and snow), adjusting the sun position, or initiating a fly-through or walkthrough
animation.

References

Blender Online Community. (2016). *Blender—A 3D modelling and rendering package*. http://www.blender.org

Felinto, D. (2015). *Virtual reality viewport*. https://github.com/dfelinto/virtual_reality_viewport. Accessed Jan 12, 2018.

Gobster, P. H., Nassauer, J. I., Daniel, T. C., & Fry, G. (2007). The shared landscape: What does aesthetics have to do with ecology? *Landscape Ecology, 22*(7), 959–972.

Nassauer, J. I. (1997). *Cultural sustainability: Aligning aesthetics and ecology* (pp. 67–83). Washington, DC: Island Press.

Open HMD Team. (2016). *Open HMD*. http://www.openhmd.net/index.php/showcase/blender-openhmd/ Accessed Jan 12, 2018.

Smith, J. W. (2015). Immersive virtual environment technology to supplement environmental perception, preference and behavior research: A review with applications. *International Journal of Environmental Research and Public Health, 12*(9), 11486–11505.

Steinitz, C. (2012). *A framework for geodesign: Changing geography by design*. Redlands, CA: Titolo collana. Esri.

Tabrizian, P., Baran, P. K., Smith, W. R., & Meentemeyer, R. K. (2018). Exploring perceived restoration potential of urban green enclosure through immersive virtual environments. *Journal of Environmental Psychology, 55*, 99–109.

Chapter 6
Basic Landscape Analysis

Tangible Landscape allows us to explore the spatial patterns of topographic parameters and their relation to basic surface geometry. We can analyze the topography of a landscape model and how it changes by continually 3D scanning the model and computing DEMs from the scanned point clouds using binning or interpolation. By computing basic topographic parameters, morphometric units, and DEM differencing we can map changes in elevation, slope, and landform as the model is modified. These maps are then projected over the physical model of the landscape so that we have near real-time feedback and can understand the impact of our changes as we make them.

6.1 Processing and Analyzing the Scanned DEM

To start working with Tangible Landscape, the setup is first calibrated as described in Sect. 2.2.3. After successful calibration the model can be scanned continuously. With each scan the edges of the model are detected and georeferenced and the point cloud is converted into a DEM via binning or interpolation. Standard GRASS GIS modules are then used to compute maps of topographic parameters or landforms that are then projected over the model. This chapter explains how the DEM is derived and demonstrates the results in a case study.

6.1.1 Creating DEM from Point Cloud

There are two ways to derive a raster DEM from a set of points in Tangible Landscape:

© The Author(s) 2018
A. Petrasova et al., *Tangible Modeling with Open Source GIS*,
https://doi.org/10.1007/978-3-319-89303-7_6

Binning A cell value is assigned based on the univariate statistics of the points that fall inside that particular cell. When constructing a DEM the mean of the z coordinates is typically used. Binning is fast, but creates a rough surface, possibly with empty cells, so it is only useful for certain types of analyses (for example for detecting objects, see Chap. 4).

Interpolation Cell values are estimated from available point data using spatial interpolation technique, namely regularized spline with tension (RST). Interpolation produces generally smoother DEM without empty cells, which is more suitable for deriving topographic parameters and modeling processes such as water flow. For more discussion on cell size selection, see Sect. 2.2.3.

6.1.2 Interpolation with the RST Function

The point clouds produced by the scanning process are usually noisy. The regularized spline with tension (RST) function, used in *r.in.kinect* and implemented as the *v.surf.rst* module in GRASS GIS, can be used to interpolate smooth DEMs from noisy point data and simultaneously compute topographic parameters such as slope and curvature. RST approximates a surface from data points by minimizing a smoothness seminorm and the deviations between the given points and the resulting surface (Mitasova et al. 2005). The RST smoothness seminorm includes derivatives of all orders with their weights decreasing with the increasing derivative order leading to the function:

$$z(\mathbf{r}) = a_1 + \sum_{j=1}^{N} \lambda_j R(\rho_j) \tag{6.1}$$

$$R(\rho_j) = -[E_1(\rho_j) + \ln(\rho_j) + C_E] \tag{6.2}$$

where $z(\mathbf{r})$ is the elevation at a point $\mathbf{r} = (x, y)$, a_1 is a trend, λ_j are coefficients, N is a number of given points, $R(\rho_j)$ is a radial basis function, $\rho_j = (\varphi r_j/2)^2$, φ is a generalized tension parameter, $r_j = |\mathbf{r} - \mathbf{r}_j|$ is a distance, $C_E = 0.577215$ is the Euler constant, and $E_1(\rho_j)$ is the exponential integral function (Abramowitz and Stegun 1965; Mitasova and Mitas 1993). The coefficients a_1, $\{\lambda_j\}$ are obtained by solving the system of linear equations:

$$\sum_{j=1}^{N} \lambda_j = 0 \tag{6.3}$$

$$a_1 + \sum_{j=1}^{N} \lambda_j \left[R(\rho_j) + \delta \frac{w_0}{w_j} \right] = z(\mathbf{r}_i), \qquad i = 1, \ldots, N \tag{6.4}$$

where w_0/w_j are positive weighting factors representing a smoothing parameter at each given point $\mathbf{r_j} = (x_j, y_j)$.

Tension influences the detail of the surface and smoothing influences the deviations between the given points and the resulting surface. A higher smoothing value can be used to reduce the noise in the data. Theoretically, the RST method requires solution of a system of N linear equations equal to the number of given points, making the method computationally intractable for large data sets. To make the method applicable to thousands and even millions of points the *v.surf.rst* implementation uses a quadtree segmentation algorithm with smooth overlaps. Depending on the distribution of the input points in relation to the surface complexity, interpolation with *v.surf.rst* may require experimental, iterative tuning of the parameters. By adjusting the parameters which control the RST function properties and distribution of points used for interpolation, this function can be used to accurately model smooth or rough topography (Mitasova and Mitas 1993; Mitasova et al. 2005). The module can be also used to compute basic topographic parameters simultaneously with interpolation.

6.1.3   Analyzing the DEM

The geometry of an elevation surface at any point can be described by topographic parameters—slope, aspect, and several types of curvatures (Mitasova et al. 2005; Olaya 2009). Elevation surface can be partitioned into units with specific geometric properties; one approach to defining these units is geomorphons (Jasiewicz and Stepinski 2013).

Basic Topographic Parameters   The steepest slope angle β in degrees or percent and the aspect angle α in degrees are the most commonly used topographic parameters. They represent the magnitude and direction of the surface gradient vector $\nabla z = (f_x, f_y)$ (its direction is upslope) and are computed as follows:

$$\beta = \arctan\sqrt{f_x^2 + f_y^2} \qquad \beta[\%] = 100\sqrt{f_x^2 + f_y^2} \qquad (6.5)$$

$$\alpha = \arctan\frac{f_y}{f_x} \qquad (6.6)$$

where $f_x = \partial z/\partial x$ and $f_y = \partial z/\partial y$ are the first order partial derivatives of elevation surface function $z = f(x, y)$. The slope values range from $0°$ to $90°$ and the aspect values range from $0°$ to $360°$. Therefore, computing the correct angle for aspect requires the evaluation of all possible combinations of negative, positive and zero values of f_x, f_y in relation to the selected direction of $0°$ (usually east or north).

For applications in geosciences the curvature in the gradient direction (profile curvature) is important because it reflects the change in the slope angle and thus controls the change of the velocity of mass flowing downwards along the slope curve. The equation for the profile curvature is

$$\kappa_s = \frac{f_{xx} f_x^2 + 2 f_{xy} f_x f_y + f_{yy} f_y^2}{p \sqrt{q^3}} \tag{6.7}$$

where κ_s is the profile curvature in m^{-1} and f_{xx}, f_{xy}, f_{yy} are the second order partial derivatives of the elevation surface function $z = f(x, y)$, $p = f_x^2 + f_y^2$ and $q = p + 1$. The curvature in a direction perpendicular to the gradient (tangential curvature) reflects the change in the aspect angle and influences the divergence/convergence of water flow. The equation for tangential curvature at a given point is

$$\kappa_t = \frac{f_{xx} f_y^2 - 2 f_{xy} f_x f_y + f_{yy} f_x^2}{p \sqrt{q}} \tag{6.8}$$

where κ_t is the tangential curvature in m^{-1}. Partial derivatives can be computed from a suitable approximation function $z = f(x, y)$ such as local second order polynomial or certain types of splines, such as RST.

Partial Derivatives from a Raster DEM We can use a second order polynomial approximation of a surface defined by a given grid point and its 3×3 neighborhood to estimate the topographic parameters (Horn 1981):

$$z(x, y) = a_0 + a_1 x + a_2 y + a_3 xy + a_4 x^2 + a_5 y^2 \tag{6.9}$$

By fitting this polynomial to the 9 grid points (the given point $z_{i,j}$ and its 3×3 neighborhood as shown below in Fig. 6.1) using weighted least squares we can derive the coefficients of this polynomial as well as its partial derivatives ($f_x = a_1$, $f_y = a_2$, $f_{xx} = 2a_4$, $f_{yy} = 2a_5$, $f_{xy} = a_3$):

$$f_x = \frac{(z_{i-1,j-1} - z_{i+1,j-1}) + 2(z_{i-1,j} - z_{i+1,j}) + (z_{i-1,j+1} - z_{i+1,j+1})}{8 \Delta x} \tag{6.10}$$

$$f_y = \frac{(z_{i-1,j-1} - z_{i-1,j+1}) + 2(z_{i,j-1} - z_{i,j+1}) + (z_{i+1,j-1} - z_{i+1,j+1})}{8 \Delta y} \tag{6.11}$$

where Δx and Δy is the resolution (grid spacing) in the east-west and north-south directions respectively. To compute the second order partial derivatives we first denote $D(i, \delta) = z_{i,j+1} + z_{i,j-1} - 2z_{i,j}$ and $D(\delta, j) = z_{i+1,j} + z_{i-1,j} - 2z_{i,j}$. Then we can write:

$$f_{xx} = \frac{D(\delta, j+1) + (4z_{i-1,j} + 4z_{i+1,j} - 8z_{i,j}) + D(\delta, j-1)}{6(\Delta x)^2} \tag{6.12}$$

Fig. 6.1 The elevation values of a grid cell and its 3×3 neighborhood

$z_{i-1,j+1}$	$z_{i,j+1}$	$z_{i+1,j+1}$
$z_{i-1,j}$	$z_{i,j}$	$z_{i+1,j}$
$z_{i-1,j-1}$	$z_{i,j-1}$	$z_{i+1,j-1}$

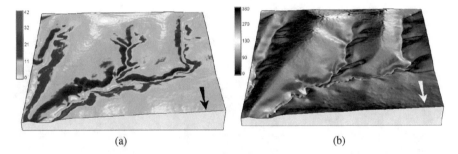

Fig. 6.2 First order topographic parameters: (**a**) slope and (**b**) aspect maps in degrees, draped over a DEM

$$f_{yy} = \frac{D(i-1, \delta) + (4z_{i,j+1} + 4z_{i,j-1} - 8z_{i,j}) + D(i+1, \delta)}{6(\Delta y)^2} \qquad (6.13)$$

$$f_{xy} = \frac{(z_{i-1,j-1} - z_{i+1,j-1}) - (z_{i-1,j+1} - z_{i+1,j+1})}{4\Delta x \Delta y} \qquad (6.14)$$

where $z_{i,j}$ is the elevation value at row j and column i, Δx is the east-west grid spacing, and Δy is the north-south grid spacing (resolution) (Figs. 6.2 and 6.3).

Landforms Landforms can be identified automatically by fitting a quadratic function to the elevation values in a given neighborhood or by "moving window" using least squares. This method for identifying landforms using differential geometry is implemented in GRASS GIS as a module named *r.param.scale*. While real-world landforms are scale-dependent and may be nested, this method can only identify landforms at a single scale based on the size of the moving window.

Geomorphons—geomorphologic phonotypes—is a novel method for identifying landforms using pattern recognition developed by Jasiewicz and Stepinski (2013) and implemented in GRASS GIS as the *r.geomorphon* add-on module. A geomor-

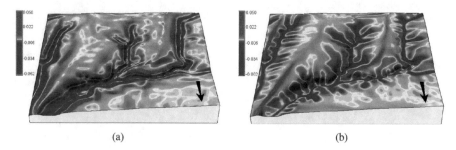

Fig. 6.3 Elevation surface curvatures: (**a**) profile and (**b**) tangential curvature maps draped over a DEM

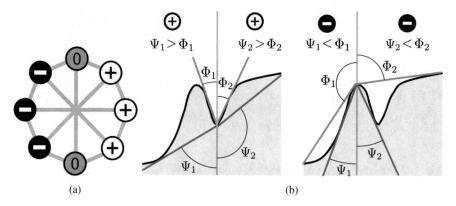

Fig. 6.4 Geomorphons use local ternary patterns and line-of-sights concepts applied to landform identification: (**a**) the relationship between a cell and its neighbors in terms of relative elevation (higher than, equal to, or lower than) is expressed as a ternary pattern; (**b**) ternary patterns are derived using the line-of-sight method and comparing the zenith angle Φ and nadir angle Ψ

phon is an abstract unit of terrain described by the local ternary pattern rather than by relief. The relationship between a cell and its neighborhood is described using an 8-tuple pattern of lower ($-$), equal (0), higher ($+$) elevation values (Fig. 6.4a). Neighborhoods are based on visibility rather than the predefined dimensions of a moving window. Geomorphons uses the line-of-sight principle to dynamically determine the optimal size of the search radius for a neighborhood based on openness of the terrain (see Fig. 6.4b). Thus, unlike methods based on differential geometry, the geomorphons approach is able to efficiently classify landforms across a range of spatial scales (Jasiewicz and Stepinski 2013) (Fig. 6.5).

Fig. 6.5 Landforms identified by geomorphons (see *r.geomorphon* manual page for the explanation of landform types)

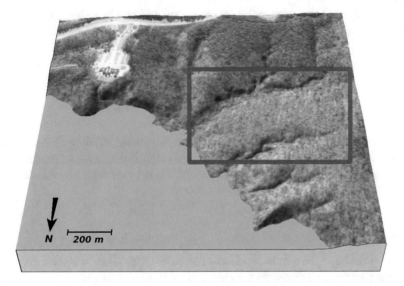

Fig. 6.6 Lake Raleigh Woods area with the development site highlighted

6.2   Case Study: Topographic Analysis of Graded Landscape

In this section we demonstrate how changes in elevation impact topographic parameters and landforms in a small study area on NCSU's Centennial Campus.

6.2.1   Site Description and 3D Model Properties

Our 56 ha case study area is located on NCSU's Centennial Campus on the southwest side of Lake Raleigh (Fig. 6.6). The topography is largely natural, but there

are some anthropogenic features, such as informal trails and the Chancellor's house
with its access road, parking, and retention pond. There is rapid development on the
campus; several new buildings are currently under construction. At the same time
there is a desire to preserve part of the campus as a park and a natural teaching
area. In this case study we developed part of the natural area as student housing and
constructed an access road. Since significant grading would be required to develop
the site given the steep slopes we extensively modified the topography and mapped
the change in elevation (cut and fill), slope, aspect, curvature, and landform.

We CNC routed the base and mold for the model from MDF based on a DEM
derived from a 2013 lidar survey. We used the base and the mold to cast a sculptable
sand model. Each side of the 1:1500 scale model is 50 cm and represents 750 m.
The model is 2 times vertically exaggerated and has 44 mm of relief which represent
33 m of elevation change.

We sculpted the sand model to grade a road and building sites. With Tangible
Landscape we were able to analyze these modifications and understand how we had
changed the slope, aspect, shape, and volume of the landscape.

6.2.2   Basic Workflow with DEM Differencing

To explore the impact of the new development we graded building sites and a road
for the future neighborhood. We start with the original molded sand model, scan it
and save the scan under different name so that it is not overwritten by the next scan.
Then we compute and display 1-meter contours:

```
g.rename raster=scan,scan_before
g.region raster=scan_before
r.contour input=scan_before output=contours_1m step=1
```

Figure 6.7a shows the terrain conditions before grading. While sculpting the model,
we can pause the continuous scanning or keep it running to get ongoing feedback
on our design as we progress.

We can visualize the magnitude and extent of our changes by differencing the
scanned elevation before and after the change and then assigning a diverging color
scheme to helps us distinguish between cut and fill (Fig. 6.7d):

```
g.rename raster=scan,scan_after
r.mapcalc "diff = scan_after - scan_before"
r.colors -n map=diff color=differences
```

Because the scans are noisy, they do not align precisely in vertical direction.
When computing difference between scans, any vertical shift is undesirable and
should be minimized. The following workflow uses linear regression to match one
scan to the other on the assumption that there has only been a moderate change in
elevation. The linear regression of the form $y = a + bx$ estimates the vertical shift a
and scale coefficient b (which should be close to 1) for adjusting the scan. Here we
define a function in Python for adjusting the scan to match the other; we will refer
to this function throughout the book:

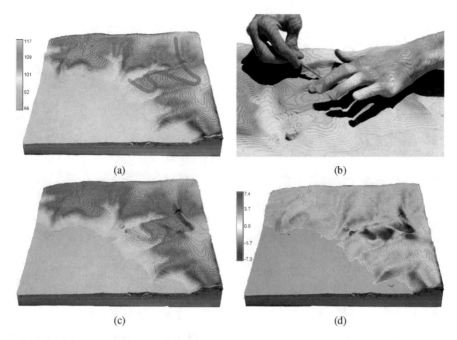

Fig. 6.7 Basic workflow: (**a**) the scanned elevation with contours, a red curve indicates the planned road; (**b**) modifying the terrain; (**c**) the finished design; (**d**) and the difference in meters between the elevation before and after grading, red representing soil removal (cut) and blue representing soil deposition (fill)

```
def adjust_scan(scan_before, scan_after, scan_adjusted):
    coeff = gscript.parse_command('r.regression.line',
        mapx=scan_after, mapy=scan_before, flags='g')
    gscript.mapcalc('{scan_adjusted} = {a} + {b} *
        {scan_after}'.format(scan_adjusted=scan_adjusted,
        a=coeff['a'], b=coeff['b'], scan_after=scan_after))
```

Here we use the function for adjusting scans to compute the difference:

```
adjust_scan('scan_before', 'scan_after', 'scan_adjusted')
gscript.mapcalc('diff = scan_adjusted - scan_before')
gscript.run_command('r.colors', flags='n', map='diff',
    color='differences')
```

6.2.3   The Impact of Model Changes on Topographic Parameters

In our case study the new neighborhood is planned on an east-facing hillside with a gentle slope of about 5° (Fig. 6.8). The slope values are steep on two sides of the hillside where it changes into parallel valleys; the profile curvature increases at the

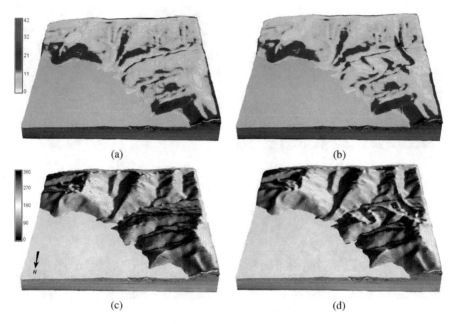

(a)   (b)

(c)   (d)

Fig. 6.8 Terrain parameters: (**a**, **b**) slope and (**c**, **d**) aspect in degrees computed from the scanned model before (**a**, **c**) and after (**b**, **d**) modification

edge of the valleys (Fig. 6.9). The profile curvature on the hillside varies between positive and negative values denoting convex (ridges) and concave shapes (valleys).

We computed the slope and aspect using the *r.slope.aspect* module based on the standard polynomial approximation using 3×3 window, see Eqs. (6.9)–(6.14). When computing curvature, we use a more flexible method with a larger window (21×21 grid cells) to capture the larger scale morphology associated with the main valleys and ridges at our study site. This method is implemented in the *r.param.scale* module and we used this module to compute profile and tangential curvature. To visualize the variability in the curvature values we used a dedicated color ramp with divergent scheme:

```
r.slope.aspect elevation=scan aspect=scan_aspect \
    slope=scan_slope
r.param.scale input=scan output=scan_pcurv size=21 method=longc
r.param.scale input=scan output=scan_tcurv size=21 method=crosc
r.colors map=scan_pcurv color=curvature
r.colors map=scan_tcurv color=curvature
```

To build a new road with gentle slopes we constructed a series of switchbacks with low areas raised on embankments. As a result there are steep slopes beside the road on the faces of the embankments (Fig. 6.8). We also flattened parts of the hillside as building sites. When we analyzed the curvature of the modified landscape the flat roads were mapped as concave areas with high curvature due to the larger neighborhood size used in the analysis (Fig. 6.9).

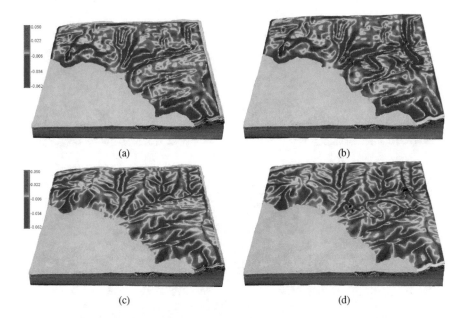

(a)   (b)

(c)   (d)

Fig. 6.9 Terrain curvatures: (**a**, **b**) profile and (**c**, **d**) tangential curvature computed from the scanned model before (**a**, **c**) and after (**b**, **d**) modification

6.2.4   Changing Landforms

The new development would change the landforms creating new ridges and valleys. We explored these changes by running the *r.geomorphon* add-on module which can be installed using *g.extension*. We can tune the results with parameters such as *search* and *skip* that determine the appropriate scale of analysis. By increasing these parameters we can skip small terrain variations in order to capture larger landforms:

```
g.extension r.geomorphon
r.geomorphon dem=elevation forms=landforms search=16 skip=6
```

In the resulting landform classification the new road has formed a major ridge (Fig. 6.10).

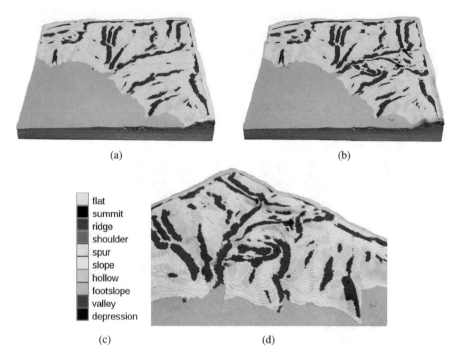

Fig. 6.10 Landform classification: (**a**) before development, (**b**) after development with the new elevated road classified as a ridge, (**c**) legend and (**d**) a detailed view of the modification done in Fig. 6.7

References

Abramowitz, M., & Stegun, I. (1965). *Handbook of mathematical functions: With formulas, graphs, and mathematical tables* (Vol. 55). New York: Dover Publications.

Horn, B. K. P. (1981). Hillshading and the reflectance map. *Proceedings of the IEEE, 69*(1), 41–47.

Jasiewicz, J., & Stepinski, T. F. (2013). Geomorphons – A pattern recognition approach to classification and mapping of landforms. *Geomorphology, 182*, 147–156.

Mitasova, H., & Mitas, L. (1993). Interpolation by regularized spline with tension: I. Theory and implementation. *Mathematical Geology, 25*(6), 641–655.

Mitasova, H., Mitas, L., & Harmon, R. (2005). Simultaneous spline approximation and topographic analysis for lidar elevation data in open-source GIS. *IEEE Geoscience and Remote Sensing Letters, 2*, 375–379.

Olaya, V. (2009). Basic land-surface parameters. In *Geomorphometry concepts, software, applications, developments in soil science* (Vol. 33, pp. 141–169). Amsterdam: Elsevier.

Chapter 7
Surface Water Flow Modeling

The topography of the Earth's surface controls the flow of water and mass over the landscape. Modifications to the surface geometry of the land redirect water and mass flows influencing ecosystems, crop growth, the built environment, and many other phenomena dependent on water. We used Tangible Landscape to explore the relationship between overland flow patterns and landscape topography by manually changing the landscape model, while getting near real-time feedback about changing flow patterns. We coupled Tangible Landscape with a sophisticated dam breach model to investigate flood scenarios after a dam breach.

7.1 Foundations in Flow Modeling

Water flow over complex terrain can be described by a bivariate form of the shallow water continuity equation. The continuity equation can be coupled with momentum (Navier-Stokes) equations to simulate flooding due to dam failure.

7.1.1 Overland Flow

For shallow water flow the spatial variation in velocity with respect to depth can be neglected and the overland water flow during a rainfall event can be approximated by the following bivariate continuity equation (Julien et al. 1995):

$$\frac{\partial h}{\partial t} + \nabla \cdot (h \, \mathbf{v}) = i_e \tag{7.1}$$

© The Author(s) 2018
A. Petrasova et al., *Tangible Modeling with Open Source GIS*,
https://doi.org/10.1007/978-3-319-89303-7_7

where:

h is the depth of overland flow in m
t is the time in s
v is the flow velocity vector $\mathbf{v} = (v_x, v_y)$ in m/s
i_e is the rainfall excess rate (rainfall − infiltration − vegetation intercept) in m/s.

If we assume that the dynamic friction slope of the water surface can be approximated by the static bare ground slope the flow velocity can be estimated by the Manning's equation:

$$\mathbf{v} = \frac{1}{n} h^{2/3} |\mathbf{s}|^{1/2} \mathbf{s_0} \tag{7.2}$$

where:

n is the Manning's coefficient expressed here with units $\mathrm{m}^{-\frac{1}{3}}\,\mathrm{s}$
s is the negative elevation gradient $-\nabla z = -(f_x, f_y)$ of the bare ground surface
$|\mathbf{s}|$ is the magnitude of elevation gradient (slope, see Sect. 6.1.3)
$\mathbf{s_0}$ is the unit vector in the flow direction (the negative elevation gradient direction).

For steady state water flow and a steady rainfall excess rate the continuity equation has the following form:

$$\partial h / \partial t = 0 \quad \longrightarrow \quad \nabla \cdot (h\, \mathbf{v}) = i_e \tag{7.3}$$

We introduce a diffusion-like term proportional to $\nabla^2[h^{5/3}]$ to incorporate the diffusive wave effects in at least an approximate way. The spatially distributed, steady overland water flow is then expressed as follows:

$$-\frac{\varepsilon}{2} \nabla^2[h^{5/3})] + \nabla \cdot (h\, \mathbf{v}) = i_e \tag{7.4}$$

where ε is a spatially variable diffusion coefficient. The diffusion term, which depends on $h^{5/3}$ instead of h, makes the equation (Eq. (7.4)) linear in the function $h^{5/3}$ which enables us to solve it by the Green's function method using a stochastic technique referred to as a path sampling method (Mitasova et al. 2004).

The path sampling method is based on the duality between the particle and field representations of spatially distributed phenomena. In this concept the density of particles in a space defines a field and vice versa, i.e. a field is represented by particles with a corresponding spatial distribution. Using this duality, processes can be modeled as the evolution of fields or the evolution of spatially distributed particles (Mitasova et al. 2004).

The flow evolution (accumulation process) can be also interpreted as an approximation of a dynamical solution for shallow water flow, in which velocity is mostly controlled by terrain slope and surface roughness rather than water depth and friction slope. Therefore the change of velocity over time at a given location is negligible. The robustness of the path sampling method enables us to simulate complex, spatially variable conditions and efficiently explore flow patterns for a wide range of landscape configurations. This method is implemented in GRASS GIS as the module *r.sim.water*.

7.1.2 Dam Breach Flooding

Surface water flow and flooding due to a dam breach are also represented by the shallow water flow equation. For this application, however, we assume that there is no rainfall and the only source is water from the reservoir. In order to ensure a realistic representation given a large mass of moving water the shallow water continuity equation:

$$\frac{\partial h}{\partial t} + \nabla \cdot (h \mathbf{v}) = 0 \qquad (7.5)$$

is coupled with the momentum (Navier-Stokes) equation of fluid motion (Cannata and Marzocchi 2012):

$$\frac{\partial h v_x}{\partial t} + \frac{\partial h v_x^2}{\partial x} + \frac{\partial h v_x v_y}{\partial y} = S_x \qquad (7.6)$$

$$\frac{\partial h v_y}{\partial t} + \frac{\partial h v_y v_x}{\partial x} + \frac{\partial h v_y^2}{\partial y} = S_y \qquad (7.7)$$

where $\mathbf{S} = (S_x, S_y)$ is the source vector (water flowing from the lake):

$$S_x = -gh \left(\frac{\partial z_w}{\partial x} + \frac{n^2 v_x |\mathbf{v}|}{h^{4/3}} \right) \qquad S_y = -gh \left(\frac{\partial z_w}{\partial y} + \frac{n^2 v_y |\mathbf{v}|}{h^{4/3}} \right) \qquad (7.8)$$

and

g is the gravitational acceleration in $\mathrm{m\,s^{-2}}$
z_w is the water level expressed as elevation above sea level in m
n is the Manning's roughness coefficient expressed here with units $\mathrm{m^{-\frac{1}{3}}\,s}$
$|\mathbf{v}|$ is the flow velocity magnitude $|\mathbf{v}| = (v_x^2 + v_y^2)^{1/2}$ in m s.

The resulting partial differential equations are hyperbolic and non-linear and therefore must be solved numerically. The finite volume method with an upwind conservative scheme proposed by Ying et al. (2004) is used to solve the equations on a regular grid making the model suitable to GIS-based implementation. Green's theorem is used to obtain the discrete equations. The numerical method explicitly solves the governing equations in two separate steps. First, the continuity equation 7.5 is evaluated deriving the water depth at time $t + \Delta t$. Then these values are used in the source term (Eq. (7.8)) to solve the momentum equations (7.6), (7.7) and evaluate the flow velocities at time $t + \Delta t$. More details about the numerical solution can be found in Cannata and Marzocchi (2012). The model is implemented in GRASS GIS as an add-on module *r.damflood*.

7.2   Case Study: The Impact of Development on Surface Water Flow

Development on North Carolina State University's Centennial Campus raises concerns about the impact of stormwater runoff during construction. We used Tangible Landscape to explore how grading potential construction sites would change overland flow and to test the design of stormwater control measures.

In this case study we worked with the same area and physical model used in Sect. 6.2.1. To familiarize ourselves with the basic hydrologic conditions of this site we computed flow accumulation and delineated watershed boundaries. First, we set the computational region to match our study area and the resolution to 1 m and then we ran the least cost path flow tracing implemented in the module *r.watershed* to derive the flow accumulation raster and watersheds. We set the threshold for the approximate size of the watershed areas to 1000 grid cells and we converted the areas to a vector representation so that we could display the watershed boundaries:

```
g.region n=223765 s=223015 e=638450 w=637700 res=1 -p
r.watershed elevation=dem accumulation=flow_accum \
    basin=watersheds threshold=1000
r.to.vect input=watersheds output=watersheds type=area
```

The resulting map that combines the flow accumulation raster with the vector representation of watershed boundaries and contours was then projected over the model in Fig. 7.1a.

To explore the potential impact of new development we graded sites for new apartment buildings with flat slopes and constructed an access road (see Fig. 6.7d). We carved a narrow culvert to allow water to flow under the road. By calculating flow accumulation and watersheds for the modified topography (Fig. 7.1b) we could immediately see that the watershed boundaries in the developed area changed significantly due primarily to the new road acting as an artificial watershed boundary.

Flow accumulation derived by the module *r.watershed* is based on the least cost path algorithm and is not designed to represent water depth and the pooling of water in depressions. Therefore, we used the shallow water flow model implemented in the module *r.sim.water* to simulate the overland flow depth during a storm event for the current conditions and after the grading. We first derived the components of elevation surface gradient $\nabla z = (f_x, f_x)$ (the parameters dx, dy in the module

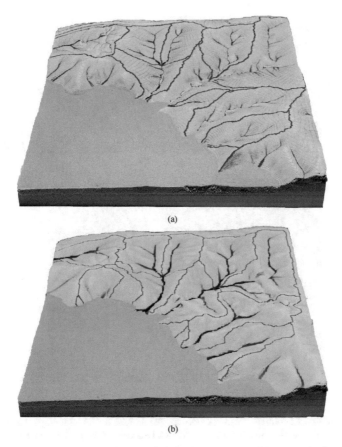

(a)

(b)

Fig. 7.1 Flow accumulation and watershed boundaries computed with *r.watershed* and projected over the sand model: (**a**) the result for the initial topography, (**b**) the flow pattern and watershed boundaries after the terrain modifications shown in Fig. 6.7

r.slope.aspect) and then ran the water flow simulation with the rainfall excess value of 150 mm/h:

```
r.slope.aspect elevation=scan dx=scan_dx dy=scan_dy
r.sim.water elevation=scan dx=scan_dx dy=scan_dy \
    rain_value=150 depth=depth
```

The resulting water depth maps were projected over the model with the initial terrain conditions and then in near real time as the terrain was modified with the final water depth pattern shown in Fig. 7.2b. Note that we have used an extremely high value of rainfall excess to provide rapid maps of water depth as the model was modified. Once we have the final design we can run the simulation with a selected storm and obtain realistic water depth estimates.

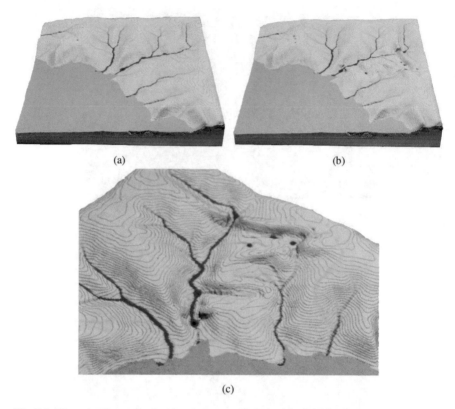

(a)   (b)

(c)

Fig. 7.2 Water depth computed with *r.sim.water*: (**a**) derived for initial topography and (**b**) after the terrain modifications shown in Fig. 6.7, (**c**) a detailed view of the modified water flow

We can also run the module *r.sim.water* in a dynamic mode (*-t* flag) and project the evolution of water depth during the storm event as an animation using the GRASS GIS animation tool.

7.3 Case Study: Dam Breach

The Lake Raleigh dam, located within our Centennial Campus study area, broke in September of 1996 after Hurricane Fran. In this case study we simulated how flood water would spread across the current landscape if the dam was breached again. We also modified the terrain to explore how different morphologies would influence the spread of flooding. We used the add-on *r.damflood*, which was designed specifically to simulate dam breaches and analyze the subsequent flooding.

Fig. 7.3 An overview of Lake Raleigh's surroundings with the dam highlighted in red. The digital elevation model on which the orthophoto is draped was created by fusing the latest 2014 data with the lake bathymetry data from 2001

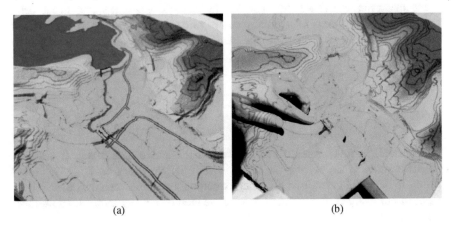

(a)   (b)

Fig. 7.4 Detailed view of the sand model in the dam location: (**a**) the current conditions with a stream and greenway below the dam; (**b**) sculpting the sand model to remove the road for the second scenario

7.3.1 Site Description and Input Data Processing

Figure 7.3 shows the terrain and the recent orthophotograph (downloaded using module *r.in.wms*) of our Lake Raleigh study area including the dam. The stream flowing out from the lake merges with the stream coming from the south in the area just below the dam, which is actively used for recreation such as disc golf, running and cycling (Fig. 7.4a).

We CNC routed a base model and a mold of the bare ground surface from MDF based on a lidar-derived DEM. The model scale is approximately 1:2400 without vertical exaggeration. We used the mold to cast a malleable layer of polymer enriched sand on top of the terrain model to create a tangible interface for exploring various dam breach flooding scenarios.

The dam breach simulation requires several input layers including the elevation raster that describes the lake bathymetry and the surrounding topography, the depth of the lake water, the Manning's roughness coefficient, and the geometry of the dam breach. The elevation raster was created by fusing the latest DEM based on the 2014 lidar survey with a DEM from the 2001 survey, which we used as a proxy for the lake bathymetry. The 2001 survey captured the topography when the lake was drained after the dam failure in 1996. We considered the 2001 DEM in the lake area a suitable approximation of the current lake bathymetry.

The lake depth raster was computed with the module *r.lake* using the fused DEM as an input. This module fills an area with water for a designated elevation level starting from a given seed point and outputs a map of the water depth for the flooded area. We determined that the current water level elevation for the Lake Raleigh was 85 m by querying the DEM along the outline of the lake. We selected an arbitrary point inside the lake as a seed point. While *r.lake* assigns null values to the cells outside of the filled area, *r.damflood* requires zero values outside of the lake, so we converted null values to zeros with the *r.null* module and limited the flooded area to the lake mask:

```
g.region n=224026 s=223007 e=639480 w=637798 res=3 -p
r.lake elevation=lake_bottom_dem water_level=85 \
    lake=lake_depth_tmp coordinates=638792,223659
r.null map=lake_depth_tmp null=0
r.mapcalc "lake_depth = if(lake_mask, lake_depth_tmp, 0)"
```

The Manning's roughness coefficient influences the speed of water. It depends on landcover and varies between 0 and 1 with higher values resulting in slower flows. To derive Manning's coefficients we could for example reclassify a landcover map. In our example for simplicity's sake we created a raster map with a uniform Manning's value:

```
r.mapcalc "manning = 0.01"
```

Finally, the module *r.damflood* requires a dam breach raster which represents the height of the breach from the top of the dam. To obtain this raster we scanned the model with the dam. Then we carved a breach into the dam and rescanned the model. We computed the difference using the function defined in code snippet in Sect. 6.2.2. We used a vertical limit of 1 m to avoid detecting differences due to noise from scanning:

```
adjust_scan('scan_before', 'scan_after', 'scan_adjusted')
gscript.mapcalc("breach = if(scan_before - scan_adjusted > 1,
    scan_before - scan_adjusted, 0)")
```

7.3.2 The Impact of the Road on Flooding

For our first flood simulation we used the current conditions. A major road with a culvert crosses the stream below the dam. We carved a culvert into the sand model by removing a narrow channel of sand from the road and scanned this model to provide input for our simulation.

Since our sand model does not incorporate the lake bathymetry we had to combine the fused DEM that includes the bathymetry with the scan. We then ran the *r.damflood* module with the input map layers and set the simulation parameters such as the length of the simulation, the time step for creating output maps, and the computational time step. Since the simulation is computationally intensive it may take several minutes to compute depending on the settings and resolution.

```
r.mapcalc "scan_combined = if(lake_mask > 0, lake_bottom_dem, \
    scan_adjusted)"
r.damflood elev=scan_combined lake=lake_depth dambreak=breach \
    manning=manning h=flood timestep=0.1 tstop=1000 deltat=10
```

The series of output raster maps generated by the simulation can be registered in the GRASS GIS Temporal Framework and then visualized using module *g.gui.animation*. The results for the current conditions are shown in Fig. 7.5. Clearly, the road creates a significant obstacle for the spread of water and causes pooling below the dam. Contained by the road the water floods upstream areas to the west.

In the next simulation we were interested in the conditions before the road was built in 2004. We simply removed the road (Fig. 7.4b) and then rescanned and recomputed the simulation (Fig. 7.6). Removing the road changes the flow pattern and as a result water spreads freely at lower depth thus impacting a larger area, but potentially reducing risk.

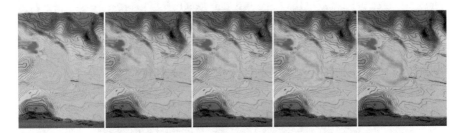

Fig. 7.5 The flood simulation with current conditions

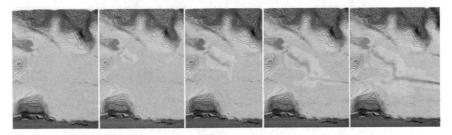

Fig. 7.6 The flood simulation after the road has been removed

7.4   Case Study: Stormwater Runoff Control Design with Flow Outside the 3D Model Area

In this case study we simulate stormwater runoff in an agricultural field and use Tangible Landscape to design runoff control measures. We use this case study to introduce modeling of processes extending beyond the 3D physical model. The boundaries of the physical model and its scale define the spatial extent of our interactions on the landscape. These boundaries often do not match the boundaries of the physical processes, such as water flow, which accumulates within watersheds. The effects of our interventions on the physical model affect water flow downstream beyond of the boundaries of the model. Similarly if water flow is modeled without considering surface runoff within the watershed, yet beyond the boundaries of the model, then the amount of water in the landscape will be underestimated. In the following case study we use smooth fusion (Petrasova et al. 2017) to combine a lidar-based DEM of entire watershed with continuous scans of a 3D physical model that is actively being modified.

7.4.1   Site Description and the Physical Model

The study area is located at the Lake Wheeler Road Field Laboratory of North Carolina State University (NCSU), Raleigh (Fig. 7.7). The area is used for agricultural research, dedicated to the production of grain crops for animal feed. In this case study, we use the physical model of the agricultural field to design stormwater runoff control measures to reduce concentrated flow causing gully erosion.

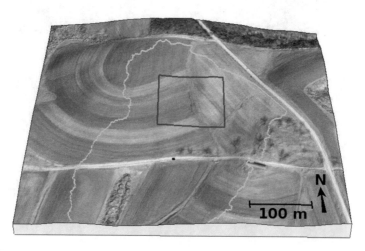

Fig. 7.7 The study area with extent of the physical model outlined in red and the watershed outlined in blue

We manually built a physical model from polymer enriched sand based on the 2015 lidar data at 1:420 scale and 4 times vertical exaggeration, to facilitate scanning and interaction. We used projected contours and the color-coded difference of the scanned and real DEM, while building the model to ensure sufficient accuracy.

7.4.2   Surface Runoff Modeling

To simulate water flow within the entire studied watershed we merged the scanned DEM of the physical model with the lidar-based DEM covering the entire watershed. Smooth fusion was essential for ensuring that the simulated water flows onto and off of the physical model (Petrasova et al. 2017). It is implemented in GRASS GIS add-on *r.patch.smooth*:

```
g.extension r.patch.smooth
```

Module *r.patch.smooth* assumes input DEMs have the same resolution and are aligned properly, therefore we first resample the lidar DEM to align with the scanned DEM. We selected a smoothing distance of 15 m to smoothly blend both DEMs.

```
g.region raster=lidar align=scan
r.resample.interp input=lidar output=lidar_resampled \
    method=bilinear
r.patch.smooth input_a=scan input_b=lidar_resampled \
    output=fused smooth_dist=15
```

We then continuously ran the water flow simulation using *r.sim.water* on the merged DEM over the watershed including the physical model. We modeled the steady state flow assuming uniform rainfall excess rate of 30 mm per hour, and uniform Manning's coefficient of 0.15. The simulation ran at resolution of 0.85 m, which is given by the resolution of the scanner multiplied by the model scale.

We then started to modify the physical model using sculpting tools and our hands to fill the actively eroding rill and divert flow to the edge of the field, while the new water flow pattern was being projected over the modified sand model. Furthermore, we built a series of checkdams to prevent erosion by reducing water flow velocity during rainstorm events. Figure 7.8 shows the simulated water flow before and after the change projected over the physical model. We can observe how water flows smoothly onto and off of the sand model due to the blending where the DEMs overlap.

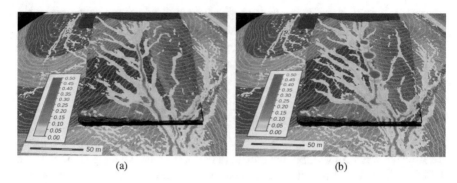

Fig. 7.8 A physical model of landscape with projected orthophoto, 20 cm contours, and simulated water flow depth in meters: (**a**) the original landscape with an eroding rill and (**b**) the landscape after our modifications

References

Cannata, M., & Marzocchi, R. (2012). Two-dimensional dam break flooding simulation: A GIS-embedded approach. *Natural Hazards, 61*(3), 1143–1159.

Julien, P. Y., Saghafian, B., & Ogden, F. L. (1995). Raster-based hydrologic modelling of spatially-varied surface runoff. *Water Resources Bulletin, 31*(3), 523–536.

Mitasova, H., Thaxton, C., Hofierka, J., McLaughlin, R., Moore, A., & Mitas, L. (2004). Path sampling method for modeling overland water flow, sediment transport, and short term terrain evolution in open source GIS. *Developments in Water Science, 55*, 1479–1490.

Petrasova, A., Mitasova, H., Petras, V., & Jeziorska, J. (2017). Fusion of high-resolution dems for water flow modeling. *Open Geospatial Data, Software and Standards, 2*(1), 6.

Ying, X., Khan, A. A., & Wang, S. S. (2004). Upwind conservative scheme for the Saint Venant equations. *Journal of hydraulic engineering, 130*(10), 977–987.

Chapter 8
Soil Erosion Modeling

Overland water flow can detach exposed soil and transport it over large distances, leading to soil loss and sediment deposition across landscape. Soil erosion can be effectively controlled by modifying topography to reduce concentrated overland flow or by planting vegetation to reduce soil detachment and transport. We used Tangible Landscape to analyze distribution of soil erosion and deposition potential in a small watershed and to design conservation measures by changing topography and planting vegetation in vulnerable locations. We iteratively adjusted and optimized our design based on real-time feedback from erosion and deposition maps projected over the modified 3D model. This feedback helped us to evaluate the effectiveness of our designs and develop better solutions.

8.1 Soil Erosion and Deposition Modeling

Soil erosion and sediment transport in landscapes is controlled by rainfall, topography, land cover, soil properties and conservation measures. It is a complex, multiscale process that is not fully understood and is challenging to predict (Jetten et al. 2003). Several models have been developed to capture and predict this process at various levels of complexity and spatial and temporal resolutions.

The *Simplified Erosion/Deposition Model* (*SEDM*) estimates sediment transport across a complex landscape and predicts the resulting pattern of erosion and deposition using the idea originally proposed by Moore and Burch (1986). The model assumes that the sediment flow rate can be approximated by sediment transport capacity and the net erosion and deposition is transport capacity limited. Under these assumptions, the net erosion and deposition can be estimated as a change in sediment flow rate along the hillslope (Mitasova et al. 2013). The model is then easy to implement using map algebra and standard flow accumulation tools available in GRASS GIS.

© The Author(s) 2018
A. Petrasova et al., *Tangible Modeling with Open Source GIS*,
https://doi.org/10.1007/978-3-319-89303-7_8

SEDM combines the Universal Soil Loss Equation parameters (Wischmeier and Smith 1978) and upslope contributing area per unit width A to estimate the sediment flow T:

$$T \approx RKCPA^m (\sin \beta)^n \tag{8.1}$$

where:

T is the sediment flow rate in $\mathrm{kg\,m^{-1}\,s^{-1}}$
R is the rainfall factor in $\mathrm{MJ\,mm\,(ha\,h)^{-1}}$
K is the soil erodibility factor in $\mathrm{ton\,h\,(MJ\,mm)^{-1}}$
A is the upslope contributing area per unit width in m
β is the steepest slope angle
C is the dimensionless land-cover factor
P is the dimensionless prevention-measures factor.

The net erosion/deposition D in $\mathrm{kg\,m^{-2}\,s^{-1}}$ is then computed as a divergence of sediment flow, equivalent to the net rate of change of the sediment mass flowing from the given grid cell (Mitasova et al. 2013):

$$D = \nabla \cdot (T s_0) = \frac{\partial (T \cos \alpha)}{\partial x} + \frac{\partial (T \sin \alpha)}{\partial y} \tag{8.2}$$

where $s_0 = (\cos \alpha, \sin \alpha)$ is the unit vector of the steepest slope direction (flow direction equivalent to the direction of negative gradient $-\nabla z$) given by the aspect angle α (see Sect. 6.1.3).

The exponents m, n in the Eq. (8.1) control the relative influence of the water and slope terms and reflect the impact of different types of flow. The typical range of values is $m = 1.0-1.6$, $n = 1.0-1.3$ with the higher values reflecting the pattern for prevailing rill erosion with more turbulent flow when erosion sharply increases with the amount of water. Lower exponent values close to $m = n = 1$ better reflect the pattern of the compounded, long term impact of both rill and sheet erosion and averaging over a long term sequence of large and small events (Mitasova et al. 2013). The relatively simple equations used in SEDM make it suitable for erosion modeling with real-time feedback when exploring erosion control alternatives with Tangible Landscape.

8.2   Case Study: Designing Erosion Control Measures

Soil erosion in agricultural areas can be controlled by modifying topography (e.g. building terraces in steep terrain) or by planting protective vegetative cover. Sediment transport from the fields can be reduced by constructing checkdams in convergent flow areas and building sedimentation ponds. We explored how various conservation practices impacted net erosion and deposition in a small watershed in the North Carolina Piedmont.

Fig. 8.1 Small agricultural watershed study site: sand model with projected orthophoto and overland water flow computed by the module *r.sim.water*

Fig. 8.2 Small agricultural watershed study site after a large storm with runoff and sediment transport in a convergent flow area

8.2.1   Site Description and 3D Model Properties

Our 25 ha case study area is a small watershed located within North Carolina State University's experimental agricultural and turf research fields (McLaughlin et al. 2001) (Fig. 8.1). Most of the watershed is used for rotating crops and turf with some areas left bare after harvest. Large storms can lead to significant runoff (Fig. 8.2) and flooding of the service road requiring mitigation and repairs.

We used lidar data to compute a bare ground DEM at 1m resolution and used it to CNC route a mold for the DEM of the study area. The physical 3D model was then cast in kinetic sand at approximately $1:2000$ horizontal scale with 3-times vertical exaggeration. The physical model served as a tangible interface for exploring how effective changes in topography and land cover are in reducing net erosion and deposition.

8.2.2   Erosion Modeling While Modifying Topography

We assumed a uniform rainfall factor $R = 4595$ MJ mm (ha h yr)$^{-1}$ as well as uniform soils and land cover with the soil erodibility factor $K = 0.02634$ ton ha h (ha MJ mm)$^{-1}$ and the land cover factor $C = 0.01$ (grass). The topographic factor exponents were set to $m = n = 1$ to represent prevailing sheet flow. Since we used uniform values for the R-, K- and C-factors, the spatial pattern of erosion and deposition depends solely upon the terrain slope and shape as well as the upslope contributing area. The resulting map represents topographic potential for erosion and deposition.

In the following workflow we first computed the slope and aspect maps of the scanned model. Then we computed the flow accumulation map and combined it with the slope and the R-, K- and C-factors using map algebra to estimate the sediment flow rate (see Eq. (8.1)).[1] Finally, we computed the divergence of the sediment flow vector field (see Eq. (8.2)) with the add-on module *r.divergence* to generate the erosion-deposition map and assigned it a custom color ramp:

```
r.slope.aspect elevation=scan aspect=scan_aspect \
    slope=scan_slope
r.flow elevation=scan flowaccumulation=scan_flowacc
r.mapcalc "scan_sedflow = 4595. * 0.02634 * 0.01 * \
    scan_flowacc * sin(scan_slope)"
r.divergence magnitude=scan_sedflow direction=scan_aspect \
    output=scan_usped
r.colors map=scan_usped rules=color_erdep.txt
```

The divergent, non-linear color ramp was specified in a plain text file `color_erdep.txt` as:

```
0% 100 0 100    #dark magenta, erosion
-100 magenta
-10 red
-1 orange
-0.1 yellow
0 200 255 200        #light green, stable
0.1 cyan
1 aqua
10 blue
100 0 0 100          #dark blue, deposition
100% black
```

---

[1]For a general case when m, n and resolution are not set to 1 see the erosion modeling tutorial: https://ncsu-geoforall-lab.github.io/erosion-modeling-tutorial/grassgis.html.

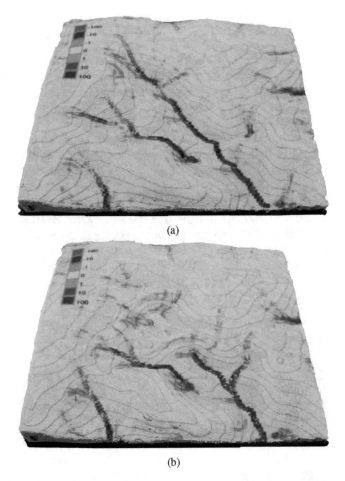

(a)

(b)

Fig. 8.3 Soil erosion and deposition maps projected over the 3D model: (**a**) result for initial terrain conditions and (**b**) after grading the terrain and creating two berms to control the impact of concentrated water flow. Yellow through red indicates the topographic potential for erosion, while light through dark blue represents sediment deposition

The resulting soil erosion and deposition map was projected over the model (Fig. 8.3). The model predicts that there is a topographic potential for high erosion due to convergent flow in the middle of the watershed and along the boundary of the agricultural field. To reduce the potential for gully formation we iteratively modified the model to create two berms in the upper part of the watershed, while observing the change in the erosion and deposition pattern projected over the model (Fig. 8.3).

8.2.3 Reducing Erosion by Modifying Land Cover

Dense vegetation cover, such as tall grasses or dense forest can reduce soil erosion and sediment transport by reducing rainfall excess and surface runoff. Vegetation intercept and higher infiltration rates limit the amount of rain that can reach the soil surface, detach it, and transport it across the landscape. These effects are represented in SEDM by the land cover C-factor (see Eq. (8.1)).

We explored the effectiveness of various configurations of protective vegetated cover using colored felt pieces to represent different shapes and types of vegetation patches and a dirt road. Spatial pattern of topographic potential for erosion and deposition, predicted by SEDM, shows the highest rates of erosion in the central valley of the watershed (Fig. 8.3). To prevent formation of a gully in this area we experimented with various designs of vegetated swales and buffers. To do this we placed felt patches over the physical model and observed the resulting erosion and deposition patterns (Fig. 8.4a). The felt patches were detected and classified based on their color into four different classes using Python function

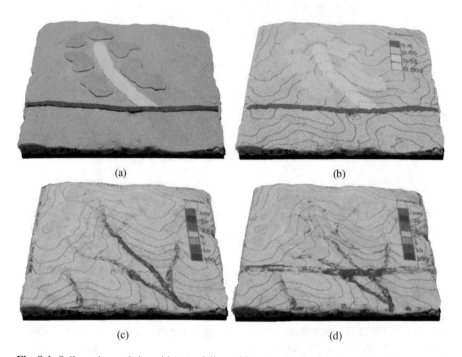

(a)   (b)

(c)   (d)

Fig. 8.4 Soil erosion and deposition modeling with protective land cover: (**a**) cutting out and placing pieces of colored felt to modify the land cover, (**b**) detecting shape and color of the felt patches and recoding to C-factor, (**c**) erosion and deposition pattern with uniform landcover, (**d**) erosion and deposition pattern with the modified, spatially variable land cover with forest, grass and a dirt road. Dark green felt (forest) and light green (grass) reduced erosion rates and increased the spatial extent of deposition. The grey felt (dirt road), however, introduced increased erosion `classify_colors` defined in Sect. 4.4. These classes are then recoded into *C*-factor classes using the following text file `recode_cfactor.txt` and module *r.recode*:

```
1:1:0.001
2:2:0.01
3:3:0.05
4:4:0.5
```

```
r.recode input=classes output=cfactor rules=recode_cfactor.txt
```

In the resulting *C*-factor raster map value 0.001 represents forest (dark green felt), 0.01 grass (light green felt), 0.05 fields (bare sand surface), and 0.5 road (grey felt). Erosion and deposition rates were then computed using the following workflow:

```
r.slope.aspect elevation=scan aspect=scan_aspect \
    slope=scan_slope
r.flow elevation=scan flowaccumulation=scan_flowacc
r.mapcalc "scan_sedflow = 4595. * 0.02634 * cfactor * \
    scan_flowacc * sin(scan_slope)"
r.divergence magnitude=scan_sedflow direction=scan_aspect \
    output=scan_usped
r.colors map=scan_usped rules=color_erdep.txt
```

The workflow is the same as in the previous section, except for the replacement of the constant *C*-factor=0.01 with a `cfactor` raster map.

References

Jetten, V., Govers, G., & Hessel, R. (2003). Erosion models: Quality of spatial predictions. *Hydrological Processes, 17*(5), 887–900.

McLaughlin, R. A., Rajbhandari, N., Hunt, W. F., Line, D. E., Sheffield, R. E., & White, N. M. (2001). The sediment and erosion control research and education facility at North Carolina State University. In *Proceedings of the International Symposium on Soil Erosion Research for the 21st Century*, Honolulu, HI, USA, 3–5 January 2001 (pp. 40–41). American Society of Agricultural and Biological Engineers.

Mitasova, H., Hofierka, J., Harmon, R., Barton, M., & Ullah, I. (2013). GIS-based soil erosion modeling. In J. Shroder & M. Bishop (Eds.), *Treatise on geomorphology, remote sensing and GIScience in geomorphology* (Vol. 3, pp. 228–258). San Diego: Academic.

Moore, I. D., & Burch, G. J. (1986). Physical basis of the length-slope factor in the universal soil loss equation. *Soil Science Society of America Journal, 50*, 1294–1298.

Wischmeier, W., & Smith, D. (1978). *Predicting rainfall erosion losses: A guide to conservation planning [USA]*. Agriculture handbook - United States. Department of Agriculture. (USA). Washington: United States Department of Agriculture.

Chapter 9
Viewshed Analysis

Viewshed (visibility) analysis is used in many different fields for both practical and aesthetic applications. It can play an important role when planning new buildings or roads especially in urban settings where obstructed views may raise safety concerns. In recreation areas views of beautiful landscapes are highly valued and protected with great passion. Visibility analysis is also crucial when planning location of monitoring cameras or communication towers in order to maximize coverage. With the increasing availability of high-resolution digital elevation models (DEMs) and digital surface models (DSMs) derived from lidar visibility analysis is becoming more accurate, broading the range of its applications. We used Tangible Landscape to analyze viewsheds on North Carolina State University's (NCSU) Centennial Campus from different observer positions and explored how future development would affect the viewsheds. We introduced object recognition to collaboratively designate observer positions.

9.1 Line of Sight Analysis

Line-of-sight analysis is used to map visible areas. If the line of sight between point A and point B does not intersect the terrain then these points are mutually visible (Fig. 9.1). More precisely, with the slope of the line between A and B defined as $(z_B - z_A)/d_{AB}$ where z is the height and d is the horizontal distance, points A and B are considered visible to each other if the line does not cross any location C such that the slope of AC is larger than the slope of AB. The viewshed of A on a raster DEM is the set of all cells of the DEM that are visible from A. The height of the observer A above terrain has significant influence on the viewshed extent. The observer height can be changed to model the view of a standing person, the view from a multi-story building, or the view from the top of a cell tower. The resulting

© The Author(s) 2018
A. Petrasova et al., *Tangible Modeling with Open Source GIS*,
https://doi.org/10.1007/978-3-319-89303-7_9

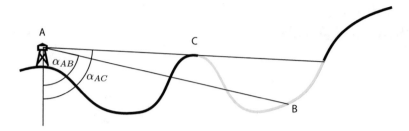

Fig. 9.1 Line of sight analysis: point B is not visible from point A because the line of sight between A and B has a smaller slope $\tan \alpha_{AB}$ than the line of sight between A and C

viewshed can be represented for example as a binary raster (containing only 0 and 1s) or a raster with the values of vertical angles with regard to the viewpoint.

Viewsheds from multiple locations can be computed and combined as the sum of binary viewsheds into a cumulative viewshed (Wheatley 1995) that allows us to assess the overall visibility of a place from selected locations. The concepts of visibility analysis have been applied to diverse geospatial applications. For example landform identification using geomorphons combines visibility analysis with computer vision (Jasiewicz and Stepinski 2013) (see Sect. 6.1.3). Several terrain relief visualization techniques take advantage of visibility analysis to highlight subtle terrain features. For example sky-view factor—the visible part of the sky unobscured by relief—can be used to render topographic relief as if it was diffusely illuminated in order to clearly visualize the relative height of features. Sky-view factor uses openness as a proxy for uniform diffuse illumination (Zakšek et al. 2011). It is implemented in GRASS GIS as the add-on module *r.skyview*. Haverkort et al. (2009) describe an efficient method implemented in the module *r.viewshed* for computing viewsheds on a raster DEM using a technique called line sweeping.

9.2 Case Study: Viewsheds Around Lake Raleigh

We used object recognition with Tangible Landscape to explore different viewsheds on NCSU's Centennial Campus by placing markers representing observer positions on a physical model of the landscape. We also evaluated how changes to the topography, the forest canopy, and the buildings would impact these viewsheds.

9.2.1 Site Description and Model

Our study area shown in Fig. 9.2 is the Lake Raleigh landscape on NCSU's Centennial Campus. Key features of the landscape include Lake Raleigh Woods

Fig. 9.2 Lake Raleigh study site: a lidar-based DSM with an orthophoto draped over the surface

in the southwest, the Chancellor's house and State Club in the south, and the Hunt Library in the northwest.

We CNC routed a base model and a mold of the landscape surface from MDF based on a lidar-derived DEM and DSM. Both models are approximately 1 : 2400 scale without vertical exaggeration. We used the base and the mold to cast a malleable layer of polymer enriched sand representing the forest canopy and buildings on top of the terrain model.

9.2.2   Visibility Analysis on DSM Using Markers

We modeled the views from different locations by placing markers on the model for Tangible Landscape to detect and recognize. Using the marker detection method described in Sect. 4.3 the position of each marker was registered as a vector point and its coordinates were then used as an input for automated computation of viewshed from this location.

We explored viewsheds from different points of interests such as the Chancellor's house, the North Shore neighborhood, the Hunt Library's observation deck, the lakeside pier, and the middle of the lake. The viewsheds were computed on the lidar-derived DSM. To identify the markers we first scanned the model and saved the scanned raster. We continued scanning while placing the markers at the locations of interest (Fig. 9.3). The markers were then identified and stored as vector points. We used the following workflow to compute the viewsheds and assign the desired color scheme. In this case visible areas are highlighted in yellow.

(a)   (b)

Fig. 9.3 Placing a marker to (**a**) identify a viewpoint and (**b**) project computed viewshed

```
coordinates = gscript.read_command('v.out.ascii',
    input='markers', separator=',').strip().splitlines()
for i, line in coordinates:
    output = 'viewshed_%s' % i
    coordinate = [float(c) for c in line.split(',')[0:2]]
    gscript.run_command('r.viewshed', flags='b',
        input='elevation', output=output,
        observer_elevation=1.75, coordinates=coordinate)
    gscript.run_command('r.null', map=output, setnull=0)
    gscript.write_command('r.colors', map=output, rules='-',
        stdin='1 yellow')
```

The viewsheds computed from different marker positions are shown in Fig. 9.4. The best views of the lake are from the Hunt Library's observation deck and the lakeside pier.

9.2.3 Modeling Viewsheds from a New Building

In the previous section we computed viewsheds on the lidar-derived DSM. In this section we used the scanned surface to model viewsheds on a modified landscape. We explored a scenario in which we built a hotel planned on the southeastern side of the lake and assessed the viewshed from the new hotel in Fig. 9.5a (the new hotel was later built in 2017, see Fig. 9.6).

We began by sculpting the new building in sand. Then we scanned and saved the modified sand surface. We placed a marker that was automatically detected and digitized as a viewpoint. We computed the viewshed using the base scan rather than the newer scan with the marker because the marker would act as an obstacle blocking the observer's view. Figure 9.5c shows the viewshed from the southwest side of the new hotel. The hotel is oriented to benefit the most from lake views. Since the hotel is a tall building it is visible from many places around the lake such as Hunt Library (Fig. 9.5d).

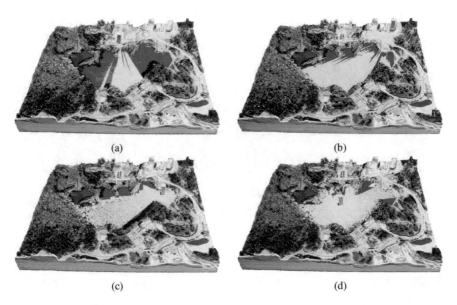

Fig. 9.4 Viewsheds computed on lidar-based DSM from viewpoints at (**a**) the North Shore neighborhood, (**b**) Hunt Library's observation deck, (**c**) a lakeside pier, (**d**) and a canoe on the lake

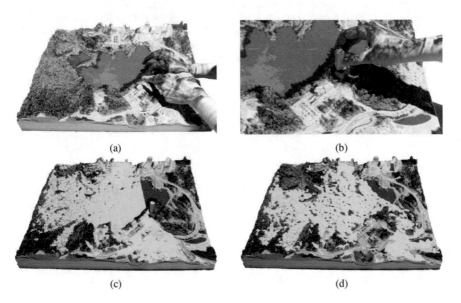

Fig. 9.5 Modeling viewsheds from a new building: (**a**) sculpting a new hotel in sand, (**b**) digitizing a viewpoint, and evaluating viewsheds from (**c**) the hotel and (**d**) the library

Fig. 9.6 View of the new hotel building from Lake Raleigh dam (December 2017)

References

Haverkort, H., Toma, L., & Zhuang, Y. (2009). Computing visibility on terrains in external memory. *Journal of Experimental Algorithmics, 13*, 5:1.5–5:1.23.

Jasiewicz, J., & Stepinski, T. F. (2013). Geomorphons – A pattern recognition approach to classification and mapping of landforms. *Geomorphology, 182*, 147–156.

Wheatley, D. (1995). Cumulative viewshed analysis: A GIS-based method for investigating intervisibility, and its archaeological application. In *Archaeology and geographical information systems: A European perspective* (pp. 171–185). London: Taylor & Francis.

Zakšek, K., Oštir, K., & Kokalj, Ž. (2011). Sky-view factor as a relief visualization technique. *Remote Sensing, 3*(2), 398–415.

Chapter 10
Trail Planning

The design of a walking or hiking trail is based on fine scale topographic conditions and varied criteria specific to the particular context such as aesthetics, views, construction cost, and environmental sensitivity. As a result trail planning is typically a product of expert knowledge, field surveys, and creative design decisions—often made on site. However, when high resolution data is available geospatial modeling can be used to identify routes optimized for travel time and suitability. To design trails with Tangible Landscape we can hand place waypoints on a physical model and then the optimal network connecting the waypoints is computed in near real-time. This approach—hand placing tangible waypoints and computationally networking the waypoints—combines creative, collaborative decision making with mathematical optimization. In this chapter we explain the theory and methodology for designing trails with Tangible Landscape and then discuss a case study, the design of hiking trail scenarios for Lake Raleigh Woods, North Carolina.

10.1 Trail Design Methodology

Our approach for designing trail networks with Tangible Landscape combines the creative identification and siting of key waypoints like trailheads and scenic spots with computationally optimized routing between these points. With Tangible Landscape we can place tangible markers by hand to automatically digitize waypoints. This allows us to work intuitively, feel the slopes and curvature with our hands, and easily collaborate.

To find the optimal routes we compute the least cost path between pairs of waypoints over the terrain and a cost surface as a function of walking energetics. First we create a cost surface that represents friction—the additional time required to cross a cell. Next we designate waypoints, points through which we want the trail to pass. For each unique pair of waypoints we use the *r.walk* module to compute the

© The Author(s) 2018
A. Petrasova et al., *Tangible Modeling with Open Source GIS*,
https://doi.org/10.1007/978-3-319-89303-7_10

anisotropic cumulative cost of walking between these points across the terrain and the friction surface based on Naismith's rule for hiking times. Then we compute the least cost path between those points across this cumulative cost surface. All of the least cost paths are combined into a network of potential routes. Finally we solve the traveling salesman problem to identify the best route across this network.

10.1.1   Least Cost Path Analysis

With least cost path analysis we can determine the most cost-effective route between a source and destinations on a cost surface. The cost surface is a raster representing cost for traversing a cell and can be derived as a function of distance, slope, land cover or other relevant criteria. To find the least cost path between locations we need to compute a cumulative cost surface where each cell contains the lowest cumulative cost of traversing the space between each cell and the specified location. We also need to generate a movement direction raster tracing the movements that created the cumulative cost surface.

First we compute the walking cost raster using Naismith's rule for walking times (Naismith 1892) with further adjustments (Aitken 1977; Langmuir 1984) of the cost for specific slope intervals. Considering only the elevation surface the cost to walk across a grid cell, expressed as time T in seconds is computed as follows:

$$T = a \cdot \Delta S + b \cdot \Delta H_u + c \cdot \Delta H_{md} + d \cdot \Delta H_{sd} \qquad (10.1)$$

where:

ΔS is the horizontal distance in m
ΔH is the height difference in m
a is the time in seconds it takes to walk for 1 m on flat surface
b is the additional walking time in seconds, per meter of elevation gain ΔH_u
       on uphill slopes
c is the additional walking time in seconds, per meter of elevation loss ΔH_{md}
       on moderate downhill slopes (this value is typically negative)
d is the additional walking time in seconds, per meter of elevation loss ΔH_{sd}
       on steep downhill slopes

Up to a specific slope value threshold, walking downhill is faster; after that it becomes more difficult and adds to the time needed to cross the cell. The slope value threshold (slope factor) derived from experiments is -0.2125 corresponding to $12°$ downslope ($\tan 12° = 0.2125$).

Taking into account land cover conditions the total cost T_{total} in seconds is estimated as a linear combination of movement and friction costs using the dimensionless weight λ:

$$T_{\text{total}} = T + \lambda \cdot F \cdot \Delta S \qquad (10.2)$$

where F is the friction in s/m. It represents the additional time in seconds that it takes to walk 1 m within a given cell due to its land cover conditions. The friction map may simply be a landcover map that has been recoded as time costs. If landcover data is not available at an appropriate resolution or if there are other important factors that need to be considered, the friction map can be created through map overlay analysis. Parameters such as soils, hydrology, roads, and existing trails can then be incorporated into the cost surface using the friction map.

We used the total cost surface given by Eq. (10.2) to compute a cumulative cost surface where each cell contains the lowest cumulative cost of traversing the space between each cell in the region and the target location. The movement direction raster is also generated tracing the movements that created the cumulative cost surface. We can then compute least cost path between any point in the study area and the target location based on the cumulative cost raster and movement directions raster.

10.1.2   Network Analysis

Using least cost path analysis it is possible to create many combinations of routes between the given waypoints. This can result in a large number of connections and potential trails. We can use network analysis to find a trail loop that goes through all given waypoints in an optimal order, avoiding connections with high costs (Fig. 10.1). The solution for this well-known optimization problem—the traveling salesman problem (TSP)[1]—is implemented in the module *v.net.salesman*, which uses a heuristic algorithm yielding a good solution in a reasonable amount of time.

10.1.3   Trail Slope Extraction

When planning trails we are typically interested in the average and maximum slope of a trail. However, the extracted values are the slope values in the direction of steepest slope rather than in the direction of the trail. To compute that, we need to

---

[1]See: https://en.wikipedia.org/wiki/Travelling_salesman_problem.

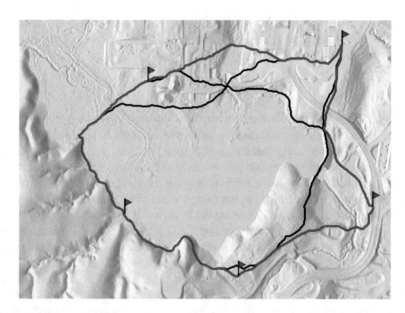

Fig. 10.1 The traveling salesman problem applied to a network of potential trails. Black lines represent potential routes. The red line is the most cost-effective combination of routes that connects all of the waypoints

know the direction of the trail, which can be derived from its vector representation. The slope along the trail β_t can then be computed as the steepest slope β (see Eq. (6.5)) multiplied by the cosine of the angle between the direction of the trail α_t and the steepest slope direction α (i.e. aspect, see Eq. (6.6)):

$$\tan \beta_t = \tan \beta \cos(\alpha - \alpha_t) \tag{10.3}$$

The difference is visible in Fig. 10.2 where part of a trail along a contour line has very low slope values along the trail, but has higher slope values in the direction of the steepest slope. Both types of slope computation can be useful depending upon our objectives. When building and maintaining a trail we may be interested in the steepness of the terrain in that area, while a hiker may be more interested in the directional slope of a trail.

10.2 Case Study: Designing a Recreational Trail

We used Tangible Landscape to design alternative trail scenarios for Lake Raleigh Woods on North Carolina State University's Centennial Campus. This old growth woodland on the south shore of Lake Raleigh is rich in biodiversity with approx-

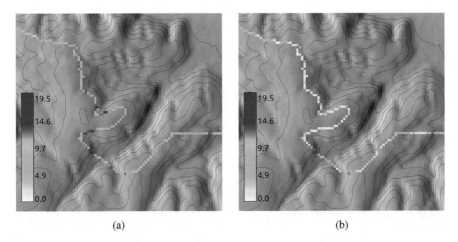

(a)   (b)

Fig. 10.2 The difference between the trail slope extracted (**a**) in the direction of the steepest terrain slope and (**b**) in the direction of the trail (in degrees)

imately 200 species of vascular plants and regionally important stands of mesic mixed hardwood and dry-mesic oak-hickory forest (Blank et al. 2010). Informal hiking and mountain biking trails have caused significant disturbance in the woods and erosion along the lakefront. A formal trail system could concentrate people on low-impact routes and thus conserve the landscape while enhancing opportunities for recreation.

10.2.1   Input Data Processing

We derived a bare earth DEM and DSM with buildings and vegetation from a lidar survey acquired in 2013. Then we CNC routed 1 : 1500 scale models of the DEM, the DSM, and their inverses (see Sect. 3.4.4). We used the inverse models as molds for casting polymeric sand models.

We created the friction map—a required input for computing the cost surface—by computing the average friction cost for each cell from a set of friction maps based on the recoded rasterized maps of buildings, hydrology, floodplains, roads, soils, and trails with *r.series* using the average function.

First we converted all of the necessary vector maps to raster. Before converting linear features we set the resolution to 3 m in order to give these features reasonable widths.

```
g.region n=224134 s=223005 w=637732 e=639085 res=1 -p

v.to.rast input=lake_raleigh output=lake_raleigh use=val
v.to.rast input=chancellors output=chancellors use=val
v.to.rast input=buildings output=buildings use=val
v.to.rast input=floodplain output=floodplain use=val
v.to.rast input=hydrology_areas output=hydrology_areas use=val
v.to.rast input=trail_areas output=trails use=val
v.to.rast input=roads output=roads use=val

g.region res=3
v.to.rast input=greenways output=greenways use=val
v.to.rast input=streets output=streets use=val
v.to.rast input=hydrology output=hydrology use=val
g.region res=1
```

In order to limit processing to publicly accessible areas we made a composite
map of inaccessible areas by combining maps of private land, buildings, and the
lake using *r.mapcalc*[2] and then created an inverted mask from this map.

```
r.mapcalc "masking = if((buildings ||| chancellors ||| \
    lake_raleigh), 1, null())"
r.mask -i raster=masking
```

Next, we recoded all maps as friction (time penalties—additional time it takes to
walk 1 m), and then overlaid the maps using *r.mapcalc* and *r.series*.

```
r.mapcalc "hydro = if(isnull(hydrology_areas ||| hydrology), \
    1, 10)"
r.mapcalc "flood = if(isnull(floodplain), 1, 10)"
r.mapcalc "paths = if(isnull(trails ||| greenways), 10, 1)"
r.mapcalc "transit = if(isnull(streets ||| roads), 1, 10)"

r.series input=flood,hydro,paths,transit,soils output=friction \
    method=average
r.colors -n map=friction color=ryg
r.mask -r
```

The resulting friction map is shown in Fig. 10.3.

10.2.2   Computing the Trail Using the Least Cost Path

In our case study we computed least cost paths between waypoints to create a
trail network. Least cost paths were computed for each combination of two points.
The following code snippet shows how such a network can be generated.For each

---

[2]Note that | | | is a symbol for logical OR which ignores NULL values and treats them as logical
false.

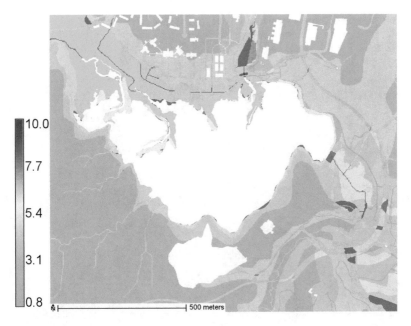

10.0

7.7

5.4

3.1

0.8

500 meters

Fig. 10.3 The friction map in seconds per meter created by combining layers of buildings, hydrology, floodplains, roads, and trails

waypoint we called the function `trail` defined below, which computes a raster representing the cumulative cost (time) from this waypoint and then multiple least cost paths from the remaining waypoints.

```
def trail(elevation, friction, walk_coeff, lambda_,
    slope_factor, point_from, points_to, vect_paths):
    # create cumulative cost surface based on walking time
    gscript.run_command('r.walk', flags='k',
        elevation=elevation, friction=friction,
        lambda_=lambda_, walk_coeff=walk_coeff,
        slope_factor=slope_factor,
        start_coordinates=point_from,
        stop_coordinates=points_to, output='tmp_walk',
        outdir='tmp_walk_dir')

    # compute least cost path to other points
    for i in range(len(points_to)):
        gscript.run_command('r.drain', flags='d',
            input='tmp_walk', direction='tmp_walk_dir',
            output='tmp_drain', drain=vect_paths[i],
            start_coordinates=points_to[i], overwrite=True)

    # remove temporary maps
    gscript.run_command('g.remove', type=['raster', 'vector'],
        name=['tmp_walk', 'tmp_walk_dir', 'tmp_drain'],
        flags='f')
```

This is repeated for all waypoints except the last.

```python
def trails_combinations(elevation, friction, walk_coeff,
    lambda_, slope_factor, points, vector_routes):
    import itertools

    coordinates = gscript.read_command('v.out.ascii',
        input=points, format='point', separator=',').strip()
    coords_list = []
    for coords in coordinates.split():
        coords_list.append(coords.split(',')[:2])

    combinations = itertools.combinations(coords_list, 2)
    combinations = [list(group) for k, group in
        itertools.groupby(combinations, key=lambda x: x[0])]

    i = k = 0
    vector_routes_list = []
    for points in combinations:
        i += 1
        point_from = ','.join(points[0][0])
        points_to = [','.join(pair[1]) for pair in points]
        vector_routes_list_drain = []
        for each in points_to:
            vector_routes_list_drain.append('route_path_' +
                str(k))
            k += 1
        vector_routes_list.extend(vector_routes_list_drain)

        trail(elevation, friction, walk_coeff, lambda_,
            slope_factor, point_from, points_to,
            vector_routes_list_drain)

    gscript.run_command('v.patch', input=vector_routes_list,
        output=vector_routes)
```

The path, however, is only computed for combinations that were not covered in previous steps in order to reduce computation time. This means that the least cost path is only computed in one direction. The direction matters because the speed of walking uphill and downhill differs. Therefore, the walking coefficients and slope factor of *r.walk* should be adjusted to affect uphill and downhill speed equally. This can be achieved by setting the slope factor to zero and using the same absolute number for coefficients b and d.

10.2.3   Finding the Optimal Trail

Using the least cost path analysis we created all of the combinations of routes between waypoints. This can result in a large number of connections and potential trails. Therefore we used network analysis to find a trail loop that goes through all our waypoints in an optimal order avoiding connections with high costs. This optimization problem—the traveling salesman problem (TSP)—can be solved using the module *v.net.salesman*.

All GRASS GIS network analysis modules require a network vector map containing connected lines (arcs) and points (nodes). This vector map can be prepared using the module *v.net* which topologically connects the given points (our waypoints) and lines (our trail combinations). Since we wanted to route the trail through all our waypoints we found the categories of all waypoints in the network and then passed them to module *v.net.salesman*:

```
def trails_salesman(trails, points, output):
    gscript.run_command('v.net', input=trails, points=points,
        output='tmp_net', operation='connect', threshold=10)
    cats = gscript.read_command('v.category', input='tmp_net',
        layer=2, option='print').strip().split()
    gscript.run_command('v.net.salesman', input='tmp_net',
        output=output, center_cats=','.join(cats), arc_layer=1,
        node_layer=2)
    # remove temporary map
    gscript.run_command('g.remove', type='vector',
        name='tmp_net', flags='f')
```

The final computation combines above defined functions and takes the scanned DEM and the detected points as input:

```
trails_combinations('scan', friction='friction',
    walk_coeff=[0.72, 6, 0, 6], lambda_=0.5, slope_factor=0,
    points='markers', vector_routes='route_net')
trails_salesman(trails='route_net', points='markers',
    output='route_salesman')
```

Examples of outputs are shown in Figs. 10.1 and 10.4.

Fig. 10.4 A trail network around Lake Raleigh

10.2.4   Mapping Trail Slopes

The simplest way to map the slope of a trail is to rasterize the trail using *v.to.rast* and then use *r.mapcalc* to extract the slope values from a slope raster computed with the module *r.slope.aspect*:

```
r.slope.aspect elevation=dem slope=slope
v.to.rast input=trail output=raster_trail type=line use=cat
r.mapcalc "trail_slope = if(raster_trail, slope)"
```

This extracts the slope values in the direction of the steepest slope. To compute the slope in the direction of the trail we first need to rasterize its vector representation by deriving raster values from the vector direction. The slope along the trail can then be computed as the slope multiplied by the cosine of the angle between the direction of the trail and the steepest slope (i.e. aspect):

```
r.slope.aspect elevation=dem slope=slope aspect=aspect
v.to.rast input=trail output=raster_trail_dir type=line use=dir
r.mapcalc "trails_slope_dir = abs(atan(tan(slope) * cos(aspect \
   - raster_trail_dir)))"
```

10.2.5   Alternative Trail Scenarios

We designed the initial trail scenario for Lake Raleigh Woods by placing markers at key points of interest—potential trailheads and scenic viewpoints (for more information about marker detection see Sect. 4.3). The siting of trailheads was informed by a previous 1 : 2500 scale study of potential routes around the entire lake (Fig. 10.4). We used realtime viewshed analysis to evaluate potential viewpoints (Fig. 10.5a). To test the viewsheds we cast the DSM in polymeric sand, carved clearings in the woodland canopy for viewpoints, placed markers to designate the viewpoints, and computed the viewsheds. We also computed the slope along the trail so that we could critique the routes (Fig. 10.5b).

Our trail needed to cross three large gullies with steep slopes. In the initial trail scenario the route follows the contours and ridge lines to avoid the gullies with their challenging slopes. This route formed two loops that intersected at the mouth of

(a)

(b)

Fig. 10.5 Trail analytics: (**a**) the viewshed from a waypoint with areas visible represented in orange, (**b**) the slope along the trail

the central gully near the lakefront with moderately steep slopes (Fig. 10.6a). We, however, wanted to experiment and see what a route with a single loop would look like so we built bridges across the gullies (Fig. 10.6b). The trail rerouted across the bridges to form a single, shorter loop with a lower average slope (Fig. 10.6c).

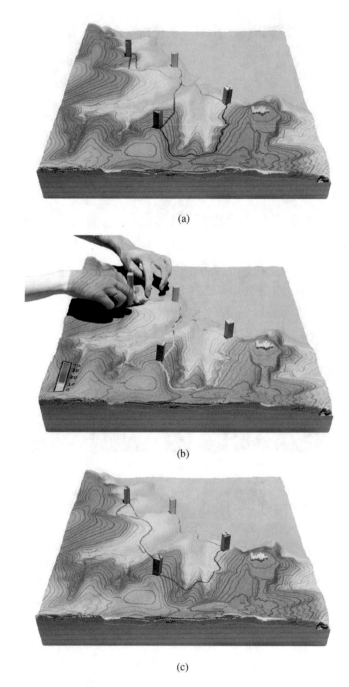

Fig. 10.6 Iteratively designing trail scenarios for Lake Raleigh Woods: (**a**) the initial route for a trail in Lake Raleigh Woods, (**b**) building bridges to reroute the trail across gullies with steep slopes, (**c**) the updated route for the trail across the new bridges

References

Aitken, R. (1977). *Wilderness Areas in Scotland*. PhD thesis, University of Aberdeen.

Blank, G., Rudder, C., Dombrowski, A., Cser, H., Lawler, M., Kollar, C., et al. (2010). *General management plan for Lake Raleigh Woods*. Technical report, College of Natural Resources, North Carolina State University, Raleigh.

Langmuir, E. (1984). *Mountaincraft and leadership*. Edinburgh: The Scottish Sports Council.

Naismith, W. W. (1892). Excursions. Cruach Ardran, Stobinian, and Ben More. *Scottish Mountaineering Club Journal, 2*(3), 136.

Chapter 11
Solar Radiation Dynamics

Solar radiation (insolation) is the primary driving force for Earth's atmospheric, biophysical, and hydrologic processes. Knowing the amount of radiation at different geographic locations at different times is therefore necessary in many fields including energy production, agriculture, meteorology, ecology, and urban planning. We modeled direct solar radiation and cast shadows in an urban setting to show how different spatial configurations of buildings change the amount of sunlight available throughout the day and the year.

11.1 Solar Radiation Modeling

Solar radiation modeling is critical for the optimal site selection of solar power plants (Carrion et al. 2008; Janke 2010) and for placing photovoltaic systems in often complex urban environments (Freitas et al. 2015; Hofierka and Kaňuk 2009; Jakubiec and Reinhart 2013). In urban design dynamic solar radiation simulations are used to study the effect of urban geometry and building configurations on solar access and shading conditions in order to get insight into the "urban canyon" phenomenon (Arnfield 1990; Lobaccaro and Frontini 2014).

On a global scale solar radiation depends spatially and temporally on the orientation of the Earth relative to the Sun. At a local scale solar radiation is influenced by topography (i.e. an elevation surface's inclination, orientation, and shadows), atmospheric conditions, and surface properties such as land cover. We can analyze the spatial distribution and dynamics of solar radiation and cast shadows using the equations that relate the sun position to the geometry of the terrain surface (Hofierka and Suri 2002). The clear-sky solar radiation model applied in this chapter is based on work undertaken for the development of the European Solar Radiation Atlas (Scharmer and Greif 2000; Page et al. 2001; Rigollier et al. 2000). We include here only the equations for direct solar radiation—beam irradiance—which

© The Author(s) 2018
A. Petrasova et al., *Tangible Modeling with Open Source GIS*,
https://doi.org/10.1007/978-3-319-89303-7_11

is influenced by the local topography. For a complete set of equations see Hofierka and Suri (2002). The diffuse and reflected components as well as atmospheric parameters can be also found in Rigollier et al. (2000).

First we compute position of the Sun in the sky as a function of time and location on Earth. The Sun declination angle depends on the day number (1 to 365 or 366):

$$\delta = \arcsin(0.3978 \sin(j' - 1.4 + 0.0355 \sin(j' - 0.0489))) \tag{11.1}$$

where δ is the Sun declination angle in radians and j' is expressed as day angle in radians:

$$j' = 360° \cdot j/365.25, \quad j = 1, 2, \ldots, 365 \, (366) \tag{11.2}$$

The position of the Sun in respect to a horizontal plane is defined by the solar altitude and solar azimuth. The solar altitude angle γ_s is a function of the solar hour angle ω, the solar declination angle δ and the latitude φ of the given location:

$$\sin \gamma_s = \cos \varphi \cos \delta \cos \omega + \sin \varphi \sin \delta \tag{11.3}$$

Given that the solar hour angle ω changes at 15° per hour and equals zero at noon it can be computed from local solar time t in decimal hours on the 24 h clock:

$$\omega = 15° \cdot (t - 12) \tag{11.4}$$

The solar azimuth angle α_s (a horizontal angle between the sun and meridian measured from east) can be expressed as a function of the solar altitude angle γ_s, the solar declination angle δ, and the latitude φ of the given location:

$$\cos \alpha_s = (\sin \varphi \sin \gamma_s - \sin \delta)/ \cos \varphi \cos \gamma_s \tag{11.5}$$

$$\sin \alpha_s = \cos \delta \sin \omega/ \cos \gamma_s \tag{11.6}$$

The position of the Sun in respect to an inclined plane with slope β and aspect α (e.g., a hillslope or a building roof) is described by a solar incidence angle v:

$$\cos v = \cos \gamma_s \sin \beta \cos(\alpha_s - \alpha) + \sin \gamma_s \cos \beta \tag{11.7}$$

To estimate the duration of solar radiation we compute the sunrise and sunset angle $\omega_{r,s}$ over a horizontal plane:

$$\cos \omega_{r,s} = - \tan \varphi \tan \delta \tag{11.8}$$

and the corresponding time of sunrise T_{sr} and sunset T_{ss} in hours is then:

$$T_{sr} = 12 - \omega_{r,s}/15° \qquad T_{ss} = 12 + \omega_{r,s}/15° \tag{11.9}$$

The sunrise/sunset angle over a south-facing inclined plane (for a location in the northern hemisphere) is computed as follows:

$$\cos \omega'_{r,s} = -\tan(\varphi - \beta)\tan\delta, \qquad \omega''_{r,s} = \min(\omega_{r,s}, \omega'_{r,s}) \qquad (11.10)$$

where $\omega''_{r,s}$ is the sunrise and sunset angle over the inclined plane. More general equations can be found in Hofierka and Suri (2002).

Solar irradiation analysis has been implemented in GRASS GIS by Hofierka and Suri (2002) in the module *r.sun*. In addition to the direct solar irradiation described above the module incorporates diffuse and reflected irradiation and cast shadows. The module *r.sun* works in two modes. In the first mode it computes the solar incident angle in degrees and solar irradiance values in $W\,m^{-2}$ for a specific instance in time. In the second mode it computes the daily sum of solar irradiation in $Wh\,m^{-2}\,day^{-1}$ and the duration of the direct irradiation. To compute a series of direct irradiance maps for a given day with a selected time interval we use the add-on *r.sun.hourly*, which calls *r.sun* in a convenient loop. Similarly we use the add-on *r.sun.daily* to compute daily sums of direct irradiation for the entire year.

11.2 Case Study: Solar Irradiation in Urban Environment

In this case study we studied how different spatial configuration of buildings changed direct solar radiation and cast shadows dynamics throughout the day. We placed wooden blocks of different shapes and sizes representing hypothetical buildings on a flat surface with an approximate scale of $1:300$. We located this abstract site north of Raleigh, North Carolina at latitude $36°$. We examined two building configurations—one with four medium-sized houses with gabled roofs and the other with three medium-sized houses with gabled roofs and one tall building with a flat roof (Fig. 11.1a, b).

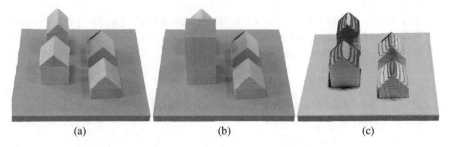

(a)   (b)   (c)

Fig. 11.1 Two configurations of buildings at approximately $1:300$ scale viewed from the south: (**a**) four medium sized houses, (**b**) one of the houses is replaced by a tall building. Figure (**c**) shows the DSM and contours derived from a scan and projected over buildings

11.2.1   The Impact of Building Configuration on Cast Shadows

We compared the shadows cast by the buildings in both configurations during summer and winter solstice. After scanning the scenes we used the add-on *r.sun.hourly* to derive binary raster time series of cast shadows from 6:00 to 22:00 (local solar time) with a 30-min time step. We had to specify the selected date as a day of year number. For example if we pick the summer solstice in 2015 we can easily compute the day of year using the Python *datetime* module:

```
import datetime
datetime.datetime(2015, 6, 21).timetuple().tm_yday
```

which gives us the number 172.

Running the following commands results in a time series of binary rasters with zeros representing cast shadow based on the direct (beam) radiation. The time series is registered in a spatio-temporal dataset and can be easily rendered as an animation and projected over the physical model.

```
r.slope.aspect elevation=scan aspect=aspect slope=slope
r.sun.hourly -t -b elevation=scan aspect=aspect slope=slope \
    start_time=6 end_time=22 time_step=0.5 day=172 year=2015 \
    beam_rad_basename=shadows_summer
g.gui.animation strds=shadows_summer
```

By changing the day parameter to 356 we obtained the time series of cast shadows for the winter solstice. Figure 11.2 compares the cast shadows at 16:30 and 19:00 (local solar time) during the summer and winter solstice for the two different building configurations.

There is a clear difference in the length of cast shadows during winter and summer. The orientation of the cast shadows at the time close to sunset changes during the year as shown in Fig. 11.2a, c. During the summer the Sun sets below the horizon later (after 19:00) far to the northwest, while it sets earlier (after 16:30) and far to the southwest during winter. As a result the house in the southeast corner only has a view of the sunset during part of the year as the summer view is blocked by another building.

By adding a tall building to the scene we cast more shadow in the space between the houses and on houses in the east. Depending upon the season one of the houses on the east side becomes shadowed by the high building sooner than the other.

11.2.2   The Impact of Building Configuration on Direct Solar Irradiation

Cast shadows give us information about the availability of direct solar radiation at a particular time on particular days. However, we are often interested in the cumulative solar irradiation, which tells us the amount of solar energy received on a given surface during a given time interval. Such information can then be used to assess the suitability of a location for solar energy applications.

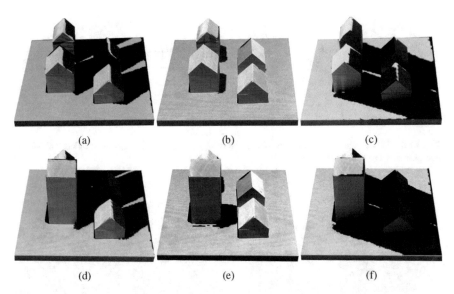

(a)   (b)   (c)

(d)   (e)   (f)

Fig. 11.2 A comparison of cast shadows viewed from south. The first and second rows show two different building configurations. The first and second columns show cast shadows at 16:30 local solar time during the winter (**a**), (**d**) and summer solstice (**b**), (**e**), respectively. The third column (**c**), (**f**) shows cast shadows at 19:00 local solar time during the summer solstice. A figure for winter solstice at 19:00 is not shown as it would be after sunset

To compute direct solar irradiation for individual days we called the add-on *r.sun.daily* with parameters similar to those we used for *r.sun.hourly*. We specified the first and last day of the year resulting in 365 daily irradiation maps. Additionally by specifying the `beam_rad` parameter we computed the sum of all of the daily maps. Again we can animate the time series projected over the scene to get more insight into the spatial patterns of solar radiation over the course of a year.

```
r.slope.aspect elevation=scan aspect=aspect slope=slope
r.sun.daily -t elevation=scan aspect=aspect slope=slope \
    start_day=1 end_day=365 day_step=1 \
    beam_rad=radiation_year_sum \
    beam_rad_basename=radiation_day_sum
g.gui.animation strds=radiation_day_sum
```

Figure 11.3 shows direct solar irradiation in $Wh\,m^{-2}\,day^{-1}$ during the winter solstice, spring equinox, and summer solstice with a common color ramp. The color ramp was specified in a plain text file `solar_color.txt` as:

```
100% red
70% yellow
0% gray
```

and then this color ramp was set for the entire time series:

```
t.rast.colors input=radiation_day_sum rules=solar_color.txt
```

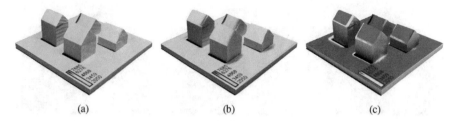

Fig. 11.3 Direct solar irradiation in Wh m^{-2} day^{-1} during: (**a**) the winter solstice, (**b**) spring equinox, and (**c**) summer solstice

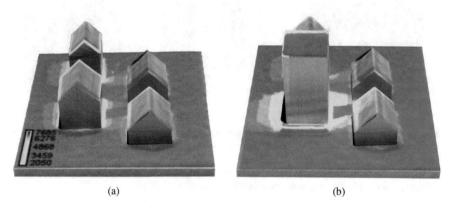

Fig. 11.4 Solar irradiation patterns in March for two different building configurations

Figure 11.4a shows the solar irradiation on a particular day in March in order to compare the cumulative solar irradiation for the two building configurations. In the first scenario there is variable irradiation in the space between the buildings as a result of the interaction between the spatial configuration of the buildings and the dynamic solar geometry. In the second scenario the pattern is even more pronounced because the street gets less direct sunlight. The slope of the buildings' roofs affects the amount of irradiation as a visual comparison of the color of the flat and tilted roofs clearly demonstrates. Furthermore, the shadow cast by the tall building has lowered the irradiation values on part of the roof of the southeastern house.

References

Arnfield, A. (1990). Street design and urban canyon solar access. *Energy and Buildings, 14*(2), 117–131.

Carrion, J. A., Estrella, A. E., Dols, F. A., Toro, M. Z., Rodríguez, M., & Ridao, A. R. (2008). Environmental decision-support systems for evaluating the carrying capacity of land areas: Optimal site selection for grid-connected photovoltaic power plants. *Renewable and Sustainable Energy Reviews, 12*(9), 2358–2380.

Freitas, S., Catita, C., Redweik, P., & Brito, M. (2015). Modelling solar potential in the urban environment: State-of-the-art review. *Renewable and Sustainable Energy Reviews, 41*, 915–931.

Hofierka, J., & Kaňuk, J. (2009). Assessment of photovoltaic potential in urban areas using open-source solar radiation tools. *Renewable Energy, 34*(10), 2206–2214.

Hofierka, J., & Suri, M. (2002). The solar radiation model for open source GIS: Implementation and applications. In *Proceedings of the Open Source GIS-GRASS Users Conference* (pp. 1–19).

Jakubiec, J. A., & Reinhart, C. F. (2013). A method for predicting city-wide electricity gains from photovoltaic panels based on LiDAR and GIS data combined with hourly Daysim simulations. *Solar Energy, 93*, 127–143.

Janke, J. R. (2010). Multicriteria GIS modeling of wind and solar farms in Colorado. *Renewable Energy, 35*(10), 2228–2234.

Lobaccaro, G., & Frontini, F. (2014). Solar energy in urban environment: How urban densification affects existing buildings. *Energy Procedia, 48*, 1559–1569.

Page, J., Albuisson, M., & Wald, L. (2001). The European solar radiation atlas: A valuable digital tool. *Solar Energy, 71*(1), 81–83.

Rigollier, C., Bauer, O., & Wald, L. (2000). On the clear sky model of the ESRA – European solar radiation atlas – With respect to the Heliosat method. *Solar Energy, 68*(1), 33–48.

Scharmer, K., & Greif, J. (2000). *The European solar radiation atlas, vol. 2: database and exploitation software.* Paris: Presses de l'Écolle des Mines.

Chapter 12
Wildfire Spread Simulation

Forest fires, whether naturally occurring or prescribed, are potential risks for ecosystems and human settlements. These risks can be managed by monitoring the weather, prescribing fires to limit available fuel, and creating firebreaks. With computer simulations we can predict and explore how fires may spread. We can explore scenarios and test different fire management strategies under different weather conditions. Using Tangible Landscape and the GRASS GIS wildfire toolset we simulated several wildfire scenarios. We tested different configurations of firebreaks on the physical model and evaluated their effectiveness.

12.1 Fire Spread Modeling Methods

Fire spread across landscapes is a complex, highly dynamic process influenced by weather, topography, and fuel. The fire spread models predict the rate of fire spread and fire perimeter growth using a combination of physical principles such as energy conservation and empirical parameters derived from observations and experimental data. For reviews of physical and empirical models of fire spread and methods of implementing fire simulation please refer to Sullivan (2009a,b,c).

12.1.1 Input Data

The fuel model is one of the most important input variables for fire simulation. Albini (1976) and Anderson (1982) describe 13 classes of fuel (Table 12.1), which differ in fuel loads and the distribution of fuel particle size classes. The size of individual pieces of fuel influences the fire; heat is absorbed faster by small twigs due to their large surface to volume ratio. The moisture content of the fuel affects

© The Author(s) 2018
A. Petrasova et al., *Tangible Modeling with Open Source GIS*,
https://doi.org/10.1007/978-3-319-89303-7_12

Table 12.1 The fuel model
by Anderson (1982) describes
13 classes of fuel

Fuel class	Description
Grass and grass-dominated	
1	Short grass (1 foot)
2	Timber (grass and understory)
3	Tall grass (2.5 feet)
Chaparral and shrub fields	
4	Chaparral (6 feet)
5	Brush (2 feet)
6	Dormant brush, hardwood slash
7	Southern rough
Timber litter	
8	Closed timber litter
9	Hardwood litter
10	Timber (litter and understory)
Slash	
11	Light logging slash
12	Medium logging slash
13	Heavy logging slash

the amount of heat needed to ignite the fuel. The smaller the fuel size, the faster the
fuel dries out and becomes more combustible.

The slope of the terrain and the speed and direction of the wind control the
rate of spread and direction of the fire. Steeper slopes cause faster ignition in the
upslope direction. This rule has been observed and quantified in several laboratory
experiments (Viegas 2004; Weise and Biging 1996; Silvani et al. 2012).

Wind is specified by speed and direction at a given height. Unlike the slope of the
terrain, wind is difficult to characterize due to its variability in space and time and
spatially and temporally averaged values are used in the models. Sparks and embers
can be carried great distances by strong winds; this process known as spotting allows
fire to spread beyond firebreaks (Albini 1983).

12.1.2   Fire Spread Algorithm

In our simulation we used the GRASS GIS modules *r.ros* and *r.spread*, implemented
by Xu (1994), to compute the rate of spread and simulate the spread of fire.

The rate of spread computation is based on the BEHAVE model (Andrews 1986).
The inputs are the fuel model, fuel moisture, wind speed and direction, and terrain
slope and aspect. The rate of spread is computed using the following equation:

$$R = \frac{I_R \xi (1 + \Phi_W + \Phi_S)}{\rho_b \varepsilon Q_{ig}} \qquad (12.1)$$

where:

R is the rate of spread in m/s
I_R is the reaction intensity in kW/m
ξ is the propagating flux ratio
Φ_W is the wind coefficient
Φ_S is the slope factor
ρ_b is the oven-dry fuel per cubic meter of fuel bed in kg/m^3
ε is the effective heating number
Q_{ig} is the heat of preignition in kJ/kg.

For a detailed description of the underlying mathematical model and an explanation of the input data refer to Rothermel (1972).

The module *r.spread* uses Huygens' principle to simulate elliptically anisotropic spread where each cell center is a potential origin of spread and the local spread is ellipsoidal (Anderson et al. 1982). The sizes and orientations of the ellipses vary by cell. Local wind and slope directions determine the orientation of ellipses. The module *r.spread* uses a specific implementation of the least cost path algorithm.

We used the GRASS GIS module *r.fire.spread* to call the modules *r.ros* and *r.spread* streamlining the fire spread computation and visualization. With this module we can use temporally variable input conditions and export the fire spread for each time step so that we can animate the simulation with the module *g.gui.animation*.

12.2 Case Study: Controlling Fire with Firebreaks

With Tangible Landscape we can intuitively create new fire management scenarios, rapidly testing how different sizes, shapes, and placements of firebreaks affect the spread of fire. To test different management strategies we simulated how a fire might spread across NCSU's Centennial Campus and tried to contain it by creating firebreaks of different sizes and alignments. In this case study we modeled a phenomenon—fire spread—that started outside the extent of the physical model, but spread towards the model (Fig. 12.1). Once the fire spread onto the model we could interact with it by creating firebreaks. Geospatial models and simulations need not be constrained to the boundary of the physical model, i.e. the extent of the tangible user interface. The processing extent can extend beyond the interactive area (see e.g., Sect. 7.4) allowing us to work across scales.

We built a malleable physical model of the tree canopy by casting polymeric sand. To build the molds for casting we interpolated a digital elevation model (DEM) and a digital surface model (DSM) from the 2013 lidar data and then CNC routed the DEM and the inverse of the DSM. Then we cast a layer of sand representing the tree canopy between the two molds. Since this layer of sand represents tree canopy and thus fuel availability we could model clearcutting simply by removing sand. Thus we were able to intuitively design firebreaks by carving into the sand layer

(a)   (b)

Fig. 12.1 The study area on Centennial Campus: (**a**) an orthophotograph with highlighted Trailwood Drive and Chancellor's House and (**b**) available fuel map projected over the table with the tangible 3D model

and reducing the fuel load. Please refer to Sect. 9.2.1 for a more detailed description of the site and model.

12.2.1   Data Preparation

To simulate fire spread we prepared several input raster layers for the module *r.ros* including the data about terrain, wind, fuel, and moisture conditions. We also specified the coordinates of location where the fire started.

Terrain Terrain slope influences the speed of spread and aspect influences the spread direction. After we set the region to our study area we derived the slope and aspect raster maps from the provided DEM:

```
g.region n=224134 s=222501 e=639326 w=637211 res=3
r.slope.aspect elevation=elevation slope=slope aspect=aspect
```

The DEM is also used as input for computing the maximum spotting distance.

Wind Wind has two components—midflame velocity and direction—and both can be spatially variable. In this case we used the prevailing wind speed and direction obtained from the nearest weather station through the State Climate Office of North Carolina.[1] The module *r.ros* requires velocity in feet per minute so the wind velocity data acquired from the State Climate Office of North Carolina must be converted from meters per second to feet per minute (to convert multiply by a factor of approximately 197). Wind direction is typically reported by the direction *from* which it originates, clockwise from the north. The module *r.ros*, however, requires the "to" direction (also clockwise from north). To simulate spatial variability we applied a

---

[1] State Climate Office of North Carolina: www.nc-climate.ncsu.edu/.

random effect, for example, using the module *r.surf.gauss* to produce a raster map
of Gaussian deviates with a specified mean and standard deviation:

```
r.surf.gauss output=wind_speed_avg mean=542 sigma=30
r.surf.gauss output=wind_dir mean=75 sigma=20
```

Fuel The module *r.ros* uses Anderson's 13 standard fire behavior fuel models.
Fuel data layers at 30 m resolution for the USA are publicly accessible via the
LANDFIRE website.[2] Most of our study site falls into fuel classes 8 and 9, i.e.
timber litter (Fig. 12.1b).

Moisture Since fuel moisture data were not readily available for our site we
generated the required, spatially variable raster maps of 1-h fuel moisture and live
fuel moisture percentage for dry conditions:

```
r.surf.gauss output=moisture_1h mean=10 sigma=5
r.surf.gauss output=moisture_live mean=20 sigma=5
```

Starting Location The starting sources of the fire are represented as raster cells
and can be created by digitizing points or importing coordinates from a file and
converting them to a raster. We provided the fire starting location in a text file with
the point coordinates (for example `638097,222934`) and converted them to a
raster with the module *v.to.rast*:

```
v.in.ascii input=source.txt output=source separator=comma
v.to.rast input=source output=source type=point use=cat
```

12.2.2   Scenario with Multiple Firebreaks

In our first scenario the fire started near Trailwood Drive and was drawn to the
northeast by a southwesterly wind. The fire jumped over the road and quickly spread
towards the Chancellor's House (Fig. 12.2).

Using the following workflow we simulated and then animated the spread of the
fire. The output raster maps generated by this workflow represent the time needed
for fire to reach each cell from the starting source in minutes. To better visually
represent the spread of the fire we used a red-yellow-gray color ramp, stretched to
the maximum value of each output raster. To prepare the suggested color ramp we
saved the color rules in a plain text file `fire_colors.txt`:

```
0% 50:50:50
60% yellow
100% red
```

---

[2]LANDFIRE: http://www.landfire.gov/.

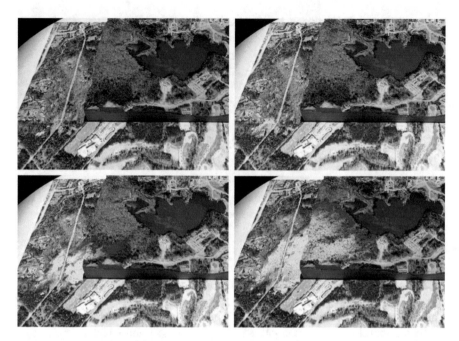

Fig. 12.2 The spread of the fire without intervention

To run the simulation we called the module *r.ros* once and then *r.spread* with the
desired time lag parameter specifying the length of the simulation. To see the
intermediate states we can run the module *r.spread* multiple times and assign our
color ramp to the resulting spread raster map:

```
r.ros model=fuel moisture_1h=moisture_1h \
    moisture_live=moisture_live velocity=wind_speed_avg \
    direction=wind_dir slope=slope aspect=aspect \
    elevation=elevation base_ros=out_base_ros \
    max_ros=out_max_ros direction_ros=out_dir_ros \
    spotting_distance=out_spotting

r.spread -s base_ros=out_dir_ros max_ros=out_max_ros \
    direction_ros=out_dir_ros start=source \
    spotting_distance=out_spotting wind_speed=wind_speed_avg \
    fuel_moisture=moisture_1h output=spread_1 lag=20
r.spread -s -i base_ros=out_dir_ros max_ros=out_max_ros \
    direction_ros=out_dir_ros start=spread_1 \
    spotting_distance=out_spotting wind_speed=wind_speed_avg \
    fuel_moisture=moisture_1h output=spread_2 lag=20
r.spread ...

r.null map=spread_1 setnull=0
r.colors map=spread_1 rules=fire_colors.txt
```

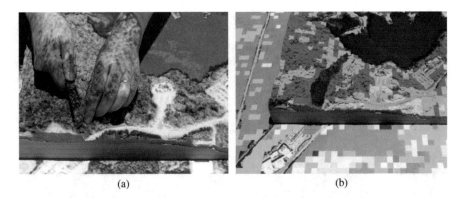

Fig. 12.3 Creating a firebreak by (a) manually removing sand and then (b) scanning and detecting the change

Alternatively we can use the module *r.fire.spread*, which conveniently wraps the previous sequence of commands into a single command:

```
r.fire.spread -s start=source times=0 end_time=1600 \
    time_step=20 output=spread model=fuel \
    moisture_1h=moisture_1h moisture_live=moisture_live \
    direction=wind_dir slope=slope aspect=aspect \
    elevation=elevation speed=wind_speed_avg
```

After simulating the initial spread of the fire we attempted to prevent fire from spreading towards Chancellor's House by creating firebreaks. First we scanned the model to save the unmodified state. Next we manually removed sand (representing canopy) from the location where we want to have a firebreak (see Fig. 12.3a). We scanned the modified model and vertically matched the new scan to the unmodified scan using the function defined in code snippet (see Sect. 6.2.2). We derived the new fuel raster layer based on the difference between the two scans introducing no data values into the copy of the original fuel model using a simple raster algebra expression:

```
adjust_scan('scan_before', 'scan_after', 'scan_adjusted')
gscript.mapcalc("changed_fuel = if(scan_before - scan_adjusted
    > 0, null(), fuel)")
```

We repeated our simulation using the new fuel raster layer. The resulting fire spread in Fig. 12.4 shows that our attempt was only a partial success. While we significantly slowed the spread of the fire, potentially giving firefighters more time to act, the fire jumped over the firebreak and started to spread towards Chancellor's House suggesting that a wider firebreak would be needed. With more precise data about fire behavior we could control and limit the spotting effect in the simulation.

In our second scenario we added another firebreak to protect the neighborhood north of Lake Raleigh. The lake creates a natural barrier that constrains the spread of the fire. To protect the community to the north we extended this barrier by creating

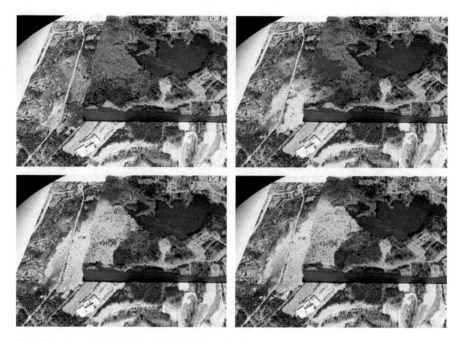

Fig. 12.4 The spread of the fire after creating a firebreak

(a)   (b)

Fig. 12.5 The simulation with additional firebreak: (**a**) the creation of a new firebreak by removing sand and (**b**) the result of the new simulation

an additional firebreak starting on western shore of the lake (Fig. 12.5a). After we rescanned the model and reran the simulation, we found that the additional firebreak was only locally effective—while the fire was significantly slowed, it would have eventually reached the neighborhood.

References

Albini, F. A. (1976). *Estimating wildfire behavior and effects*. Technical report, Intermountain Forest and Range Experiment Station, Forest Service, U.S. Department of Agriculture.

Albini, F. A. (1983). *Potential spotting distance from wind-driven surface fires*. Technical report, Intermountain Forest and Range Experiment Station, Forest Service, U.S. Department of Agriculture.

Anderson, D. H., Catchpole, E. A., De Mestre, N. J., & Parkes, T. (1982). Modelling the spread of grass fires. *The Journal of the Australian Mathematical Society. Series B. Applied Mathematics, 23*, 451–466.

Anderson, H. E. (1982). *Aids to determining fuel models for estimating fire behavior*. Technical report, Intermountain Forest and Range Experiment Station, Forest Service, U.S. Department of Agriculture.

Andrews, P. L. (1986). *BEHAVE: Fire behavior prediction and fuel modeling system – BURN subsystem, Part 1* (p. 130). Ogden: USFS.

Rothermel, R. C. (1972). *A mathematical model for predicting fire spread in wildland fuels*. Technical report, Intermountain Forest and Range Experiment Station, Forest Service, U.S. Department of Agriculture.

Silvani, X., Morandini, F., & Dupuy, J.-L. (2012). Effects of slope on fire spread observed through video images and multiple-point thermal measurements. *Experimental Thermal and Fluid Science, 41*, 99–111.

Sullivan, A. L. (2009a). Wildland surface fire spread modelling, 1990–2007. 1: Physical and quasi-physical models. *International Journal of Wildland Fire, 18*(4), 349–368.

Sullivan, A. L. (2009b). Wildland surface fire spread modelling, 1990–2007. 2: Empirical and quasi-empirical models. *International Journal of Wildland Fire, 18*(4), 369–386.

Sullivan, A. L. (2009c). Wildland surface fire spread modelling, 1990–2007. 3: Simulation and mathematical analogue models. *International Journal of Wildland Fire, 18*(4), 387–403.

Viegas, D. X. (2004). Slope and wind effects on fire propagation. *International Journal of Wildland Fire, 13*(2), 143–156.

Weise, D. R., & Biging, G. S. (1996). Effects of wind velocity and slope on flame properties. *Canadian Journal of Forest Research, 26*(10), 1849–1858.

Xu, J. (1994). *Simulating the Spread of Wildfires Using a Geographic Information System and Remote Sensing*. PhD dissertation, Rutgers University, New Brunswick.

Chapter 13
Coastal Modeling

In this chapter we discuss how to simulate inundation and flooding in coastal landscapes, explain workflows for developing storm surge and sea level rise scenarios, and present two case studies. The first case study explores storm surge and dune breach impacts for two populated barrier islands. One of these studies is designed as a simple educational game. The second study is a design—informed by flood and erosion modeling—for resilient, sustainable guest housing at a coastal research institute.

13.1 Modeling Potential Inundation

Severe precipitation, storm surge and sea level rise often cause extensive inundation and flooding of coastal landscapes. This complex process depends upon topography and bathymetry, tides, sea level surge, waves, wind, and rainfall. The spatial extent of flooding caused by storm surge can be approximated by "spreading" the observed or predicted water level throughout the landscape (Clinch et al. 2012). We can spread the given water level from predefined seed(s) located in the ocean using a 3 × 3 moving window to find all grid cells that are below the specified elevation (i.e., water level) and are connected with the seed(s) location. This approach respects the terrain barriers, so if a foredune protects a low lying area, this area will not be flooded even if the water level exceeds its elevation. This simplified algorithm assumes that the water spreads at an infinite speed and thus instantly floods everything that it can reach. Therefore, this method is best used to map the maximum possible extent of inundation assuming that there are no changes in topography during the process.

© The Author(s) 2018
A. Petrasova et al., *Tangible Modeling with Open Source GIS*,
https://doi.org/10.1007/978-3-319-89303-7_13

13.2   Case Study: Simulating Barrier Islands Flooding

We explored the impact of topography on inundation and flooding at two barrier islands in North Carolina: the Jockey's Ridge sand dune complex in Nags Head on the Outer Banks and at Bald Head Island in the southern region of the state coast. We used molds to build 3D sand models of the study areas. Then we carved breaches into the foredunes of our sand models and ran the storm surge simulations to identify which homes in these two areas were most vulnerable to flooding.

13.2.1   Storm Surge Flooding at Jockey's Ridge Sand Dunes

The study area stretched across the entire width of the barrier island and included the beach, foredune, numerous buildings, roads, large sand dunes, maritime forest, and soundside shore (Fig. 13.1). We derived a 1 m resolution bare earth digital elevation model (DEM) and a digital surface model (DSM) with buildings and vegetation for this area from a 2009 lidar survey. Then we cast a polymeric sand model of the DEM from a CNC routed mold at approximately 1 : 2000 scale. We projected an orthophotograph (downloaded using module *r.in.wms*), elevation data, the shoreline, and building footprints to provide context. Based on the building footprints we added massing models of the buildings to our sand model.

To study the potential impacts of a hurricane we simulated storm surge at increasing levels from 1 m to 4 m using the add-on module *r.lake.series*. We breached the foredune at a point where it has been compromised by the construction

Fig. 13.1 The Jockey's Ridge dune complex in Nags Head, North Carolina

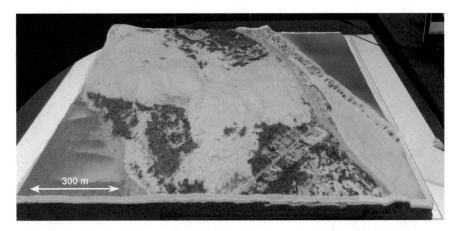

Fig. 13.2 Sand model of the Jockey's Ridge area with projected orthophoto and predicted flooding during storm surge with a foredune breach

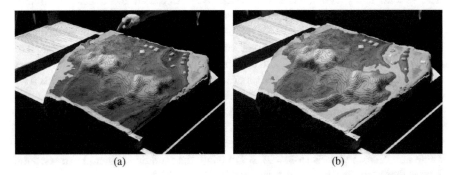

(a)   (b)

Fig. 13.3 Flooding simulations for one of the explored future scenarios at the Jockey's Ridge area for (**a**) 1 m and (**b**) 2 m storm surge water level (as measured above the 0 m vertical datum)

of a boardwalk for beach access and compared the extent of flooding at different surge levels (Fig. 13.2). Then we explored various future development scenarios, including adding new buildings, raising road elevation and additional dune construction while adjusting the scenarios based on the projected flooding extent feedback (Fig. 13.3).

```
r.lake.series elevation=elev_2009_1m output=flood \
    start_water_level=1 end_water_level=4 water_level_step=0.1 \
    coordinates=913525,249507
g.gui.animation strds=flood
```

Fig. 13.4 Sand model of a beach, homes and dunes at a Bald Head Island community used in our case study. Orthophoto is projected over the sand model

13.2.2   Exploring Storm Surge Protection

The second study area included a 700 m long stretch of the beach, foredunes, homes, neighborhood roads, coastal dunes, and maritime forest in Bald Head Island (Fig. 13.4). We derived a 1 m resolution bare earth DEM and a DSM with buildings and vegetation from a 2014 lidar survey and then we cast a polymeric sand model of the DEM from a CNC routed mold at approximately 1 : 1000 scale with four times vertical exaggeration. We projected an orthophotograph and building footprints over the model to provide context for the simulations.

To assess vulnerability of homes to flooding caused by foredune breaches we removed sand from foredunes at various locations, based on the presence of boardwalks, access roads or observed dune degradation. Then we simulated storm surge at increasing levels from 1 m to 4 m using the module *r.lake* and extracted the homes that were flooded at the simulated level (Fig. 13.5).

The following Python code snippet computes the flooding for 3 m storm surge and then it extracts buildings that were flooded. We first use raster representation of the map layer `buildings` to identify the categories of those that were flooded and then we use the vector map layer `buildings` to extract the footprints so that they can be projected in a different color on the model surface (Fig. 13.5b):

```
level = 3 # storm surge level in map units
coordinates = [703760, 11470] # seed coordinates for r.lake
gscript.run_command('r.lake', elevation='scan',
    water_level=level, lake='flood')
# extract flooded homes
lines = gscript.read_command('r.univar', flags='t',
    map='flood', zones='buildings').strip()
if lines:
    cats = [l.split('|')[0] for l in lines.splitlines()[1:]]
    gscript.run_command('v.extract', input='buildings',
        output='flooded_homes', flags='t', cats=','.join(cats))
```

(a)   (b)

Fig. 13.5   Coastal game: (**a**) adding sand to protect the homes, (**b**) simulated flood with 57 flooded homes

Fig. 13.6   Coastal storm surge educational game: successful solution designed by game participants

We have implemented this case study as a simple educational game, where the participants have a limited amount of sand and time to build protection against the storm surge assuming that the foredunes are relatively weak and can breach at any location. There is not enough sand available to completely rebuild the dunes and/or widen and raise the beach, therefore a smart strategy is needed to protect at least some homes. The breach location is then randomly generated (blue arrow in Fig. 13.5b), the model is modified at this location, while it is being scanned and a new DEM with the designed protections and a breach is computed. The flooding at a given water level is simulated and the result is projected over the model along with the footprints of homes, with the flooded homes highlighted in red (Fig. 13.5b). One of the successful solutions for a given breach created by a participant at a public event takes advantage of existing topography (Fig. 13.6).

13.3   Case Study: Designing Resilient Coastal Architecture

Two architecture students—David Koontz and Faustine Pastor—used Tangible Landscape to site a series of residential pavilions (Fig. 13.7) for visitors staying at the UNC Coastal Studies Institute on Roanoke Island, North Carolina as part

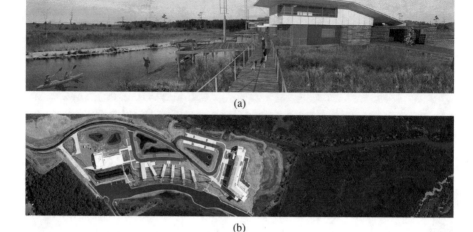

Fig. 13.7 The students' design for a series of pavilions for the UNC Coastal Studies Institute on Roanoke Island: (a) a rendering of the proposed design for a residential pavilion, (b) a masterplan of the UNC Coastal Studies Institute with the proposed residential pavilions. Images by David Koontz and Faustine Pastor

of a joint architecture-engineering studio. They used storm surge and water flow simulations to inform the selection of building sites. Then they spaced pilings for the buildings and adaptively graded a biofiltration swale based on near real-time feedback about water flow, sediment flux, and erosion-deposition.[1]

The UNC Coastal Studies Institute—the site for the studio—is a research institute on Roanoke Island in the Outer Banks of North Carolina. Students in this North Carolina State University studio course led by Professors Joe DeCarolis, David Hill, and Ranji Ranjithan were tasked with designing sustainable housing for guests and researchers visiting the institute.[2] The site is at risk of inundation from sea level rise and storm surge and is in a sensitive back-barrier salt marsh.

Given the risk of flooding David and Faustine used Tangible Landscape to visualize different levels of inundation and test the vulnerability of potential building sites. They placed basswood massing models of buildings on a cast sand model of the landscape and then used the module *r.lake* to simulate 4 m storm surge and check if their buildings would stay dry (Fig. 13.8). In their design the buildings would be raised on piers and connected by boardwalks over a bioswale. They simulated water flow to find the right spacing for the piers over swale. Then they adaptively graded the swale using feedback from water flow, erosion-deposition, and sediment

---

[1] Watch the video on Youtube: https://youtu.be/PbbzWymGvLo.

[2] Read more about the studio on the Coastal Dynamics Design Lab website: http://design.ncsu.edu/coastal-dynamics-design-lab/.

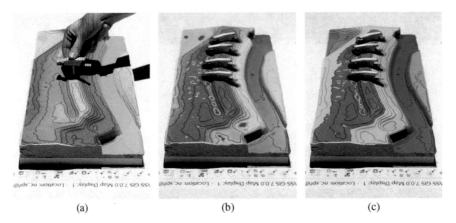

(a)   (b)   (c)

Fig. 13.8 Students using water flow and flood simulations to adaptively inform the placement of a series of pavilions for the Coastal Studies Institute on Roanoke Island, NC: (**a**) siting buildings by hand on an artificial ridge, (**b**) studying the flow of water around the buildings and (**c**) testing whether the buildings stay dry with 4 m storm surge

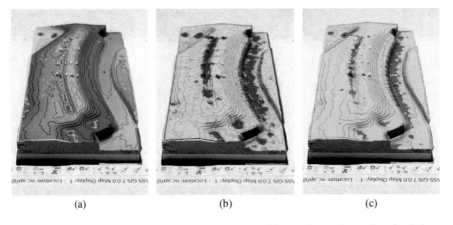

(a)   (b)   (c)

Fig. 13.9 Students using (**a**) water flow, (**b**) erosion-deposition, and (**c**) sediment flux simulations to space structural piers and adaptively grade a biofiltration swale for the Coastal Studies Institute on Roanoke Island, NC

flux simulations, sculpting the channel so that water and sediment flowed towards a bioretention pond (Fig. 13.9).

Reference

Clinch, A. S., Russ, E. R., Oliver, R. C., Mitasova, H., & Overton, M. F. (2012). Hurricane Irene and the Pea Island breach: prestorm site characterization and storm surge estimation using geospatial technologies. *Shore & Beach, 80*(2), 38–46.

Chapter 14
Landscape Design

In this chapter we demonstrate how tangible geospatial modeling can be coupled with virtual reality as a tangible immersive environment for landscape design. In a tangible immersive environment spatial scientists and landscape architects can rapidly, collaboratively design new landscapes balancing aesthetic and environmental factors. As a case study we use Tangible Landscape to create different scenarios for a park with landforms, hydrological systems, planting, a trail network, and a shelter. Real-time immersive visualizations rendered in Blender help us to judge the aesthetics of the park and refine our designs. In this case study we synthesize the concepts and workflows introduced in previous chapters, including Chaps. 4, 5, 7 and 10.

14.1 Integrating Tangible and 3D Modeling Methods

Designing a park is a complex process in which many functional and aesthetic aspects must be considered and evaluated by experts from different domains (Molnar 2015). In this chapter we demonstrate how a tangible immersive environment can improve the design process through intuitive, real-time tangible interactions and photo-realistic renderings that help experts collaborate, experiment, assess their work, and make tradeoffs. By integrating the tangible and 3D modeling methods described in previous chapters in a single case study we can experiment and rapidly create scenarios for a new park.

The system setup in Fig. 14.1 has several components allowing for different types of tangible interaction—sand model for hand sculpting, pieces of colored felt for planting, markers for designing a trail, marker for siting a shelter, and direction marker for exploring views. The setup also provides feedback about the design quantitative and aesthetic properties through the projection of map layers on the

© The Author(s) 2018
A. Petrasova et al., *Tangible Modeling with Open Source GIS*,
https://doi.org/10.1007/978-3-319-89303-7_14

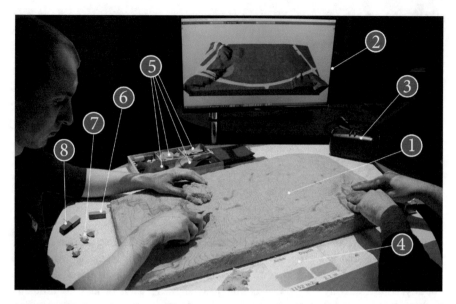

Fig. 14.1 For the design case study the Tangible Landscape setup included: 1) a hand sculpted model of the landscape, 2) a computer display showing the 3D model in Blender's viewport, 3) an Oculus DK2, 4) a projected dashboard with quantitative feedback, 5) pieces of colored felt representing patches of vegetation, 6) a marker for siting the shelter, 7) markers for siting trail waypoints, 8) and a direction maker for siting viewpoints

model and a simple dashboard on the table, a computer display with 3D model rendered in Blender, and a head-mounted display.

The design process consisted of several stages—modifying topography, planting vegetation, placing a picnic shelter, and designing a trail to the shelter (Table 14.1). At each stage we used different interaction methods with different set of analyses, which were prepared in advance using Tangible Landscape *Activities* functionality (see Sect. 2.2.4 and GitHub wiki page[1]). The code and data for running similar landscape design study is available online.[2]

Modifying Topography In the first step—sculpting the surface of the model as topography—our feedbacks were projected contours and the pattern of overland flow (see Chap. 7). Lakes were modeled separately as filled terrain depressions using the module *r.fill.dir*, which is normally used to create a depressionless DEM for flow routing algorithms. By subtracting the original scan from the filled DEM, we obtain the filled depressions representing lakes:

```
r.fill.dir input=scan output=output direction=tmp_dir
# extract depressions, filter shallow depressions
r.mapcalc "ponds = if(output-scan > 0.2, output-scan, null())"
r.colors map=ponds color=water
```

---

[1] https://github.com/tangible-landscape/grass-tangible-landscape/wiki/Working-with-Activities.

[2] https://github.com/tangible-landscape/tangible-landscape-applications/tree/master/planting.

Table 14.1 Overview of design stages and their associated interaction modes, feedback types, analyses in GRASS GIS and relevant sections of this book

Mode	Interaction	Feedback	Analysis	Section
Topography	Scultping	Projection on the model, hydrology dashboard, 3D rendering	*r.contour, r.sim.water, r.fill.dir*	7.2
Planting	Colored felt	Biodiversity dashboard, 3D rendering	*r.li*	4.4
Shelter	Markers	3D rendering		4.3
Trail	Markers	Projection of slope and profile, 3D rendering	*r.slope.aspect, r.profile*	10.2.4
3D views	Direction marker	3D rendering, head-mounted display		4.6

We computed the area and average depth of the lake on-the-fly and these values were visualized graphically on a projected dashboard (Fig. 14.1). As described in Sect. 5.5.1 the new topography and lakes were exported as GeoTIFFs to Blender where they were rendered in a viewport with predefined materials (Sect. 5.5.4).

Planting Using interaction method described in Sect. 4.4 (and specifically function classify_colors) we used pieces of colored felt to plant trees and shrubs. In the calibration phase we matched four colors to tree species—with red representing red oaks, orange representing shrubs, green representing pines, and blue representing willows. After cutting out pieces of felt and detecting them, we created patches of rectangular meshes for each tree species that are then draped over the scanned terrain and exported to Blender as 3D Shapefiles (Sect. 5.5.1). In this way the bottom of the rendered trees in the 3D model align with the topography.

```
classify_colors(new='patches', group=color)
# find out which categories of species are detected
cats = gscript.read_command('r.describe', map='patches',
    flags='1ni').strip()
for cat in cats.splitlines():
    # mask each category
    gscript.run_command('r.mask', raster='patches',
        maskcats=cat, overwrite=True)
    # create a regular 2D vector mesh within the masked area
    gscript.run_command('r.to.vect',
        input='scanned_topography', output=cat + '_2d',
        type='area', flags='svt')
    # drape 2D mesh over scanned topography to get 3D mesh
    gscript.run_command('v.drape', input=cat + '_2d',
        output=cat + '_3d', elevation='scanned_topography')
    gscript.run_command('v.out.ogr', input=cat + '_3d',
        output=cat + '_3d.shp', format='ESRI_Shapefile',
        lco='SHPT=POLYGONZ')
gscript.run_command('r.mask', flags='r')
```

In Blender, we used particle systems (explained in Sect. 5.5.2) to visualize the patch as a grove of the given tree species with randomly distributed trees. Similarly, we used colored markers to represent individual trees on the landscape.

To assess landscape structure of our scenarios we computed several biodiversity metrics that quantitatively describe the pattern, distribution, and shape of vegetation patches in the landscape. These metrics were projected as a dashboard of charts below the model as additional feedback (Fig. 14.4a). The landscape metrics include number of patches, mean shape index, Shannon's diversity index, and shape index (Baker and Cai 1992), and are implemented in a set of GRASS GIS modules named *r.li*.

Siting a Shelter To site a shelter we used a colored wooden block, and detected its position based on the difference between a saved scan before placing the shelter and the current scan (Sect. 4.3). A 3D model of a shelter is then rendered in Blender at that position.

Designing a Trail We designed a trail between two defined locations (park entrance and shelter) by placing markers representing waypoints along the trail. We used a technique defined in Chap. 10, but modified it to first calculate the optimal order of marked points from the entrance to the shelter based on their euclidean distance. This is accomplished with the TSP-solving algorithm by defining the distance between the start and end point as 0 and then removing that segment of the trail. After the order is determined, the least cost path is computed between successive pairs of points using modules *r.walk* and *r.drain*. Then the slope along the complete trail is derived (see Sect. 10.2.4), the vertical profile of the trail is computed with module *r.profile* and plotted on a dashboard, and the route is 3D rendered as a boardwalk in Blender.

3D Rendering of Views At each stage of the design process we can explore the views on the landscape from human perspective by placing and orienting the direction marker. When the direction marker is detected (Sect. 4.6) we export it as a line to Blender, which interprets it as a change in camera and renders the new view on the screen.

The real-time viewport rendering used throughout the process has a sufficient degree of realism for rapid decision-making (Sect. 5.5.3). At any point in the design process, however, we can compute full renderings for selected views of the landscape. These full, high-resolution renderings are more photorealistic and help us to make well-considered judgments about the aesthetics. They are also useful for presenting the final design scenarios to others. Additionally, we can view the resulting landscape from human perspective using head-mounted displays.

Fig. 14.2 Study area with the highlighted physical model extent, vertically exaggerated twice

14.2 Case Study: Designing a Park

Using the above described workflows we designed two scenarios of future park close to NCSU campus and compared them using the metrics, site plans, and perspective views.

14.2.1 Site Description and Model

Our 7.7 ha case study area shown in Fig. 14.2 is located between NCSU Centennial Campus and Dorothea Dix Park, and is mostly covered by dense shrubs, making the area inaccessible for any recreational activities. For our case study, we assumed that the local vegetation would be completely removed and replaced with select species of trees and shrubs. The two main roads that cross our study area—the multi-lane Centennial Parkway to the west and the low-traffic Blair Road to the south—provide the main access to Dix park from the NC State campus.

We hand sculpted the physical model with the aid of projected contours and the dynamically computed, color-coded difference (Sect. 3.1). Given the local topography and model's scale of 1 : 500, we vertically exaggerated the model twice to simplify physical interaction.

14.2.2 Scenario 1

In the first scenario we used the existing depression in south-west part of our area to create an approximately 0.5-ha lake (Fig. 14.3a). Using the excavated soil, we raised

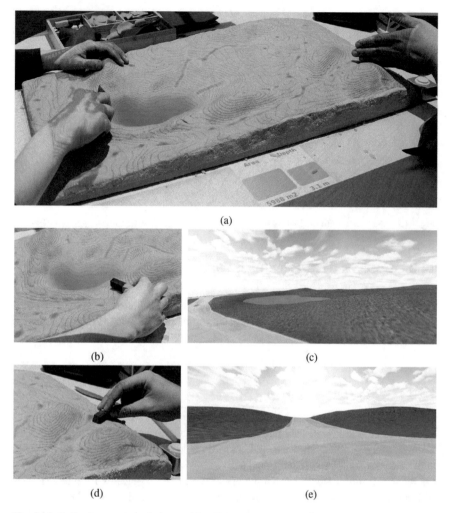

Fig. 14.3 In the first scenario designers (**a**) collaboratively sculpted the landscape to create ponds and artificial mounds. Then, they (**b**), (**d**) used the wooden marker to (**c**), (**e**) explore various views

the west banks of the lake to create a protective barrier to reduce noise coming from Centennial Parkway. The resulting undulating topography along Blair Road (Fig. 14.3a)—crossing our study area from west to east—provides interesting views for visitors (Fig. 14.3c and e). We preserved the drainage ditch along Blair Road and channeled overland water flow into the lake.

Next we planted trees and shrubs, while considering the location of the shelter and the entrance to the park on the east side. We planted red oaks and shrubs sparsely along Blair Road to buffer the park from the road (Fig. 14.4a), while offering pleasant, welcoming views to the visitors arriving at the entrance. With blue markers we planted individual willow trees along the banks of the lake. We also used cut-out patches of felt to create denser, continuously vegetated planting areas.

(a)

(b)

(c)

(d)

(e)

(f)

(g)

(h)

Fig. 14.4 In the first scenario the designers (**a**) cut out and placed felt pieces on the tangible model to (**b**) develop planting strategies, which they evaluated with the biodiversity metrics shown on the dashboard. Then, (**c, d**) they sited the shelter and (**e**) placed wooden markers to (**f**) route the boardwalk. Finally they (**g**) used the direction marker to generate (**h**) 3D renderings of various viewpoints

After several design iterations (Fig. 14.4c), we placed the shelter on the north shore of the lake with unobstructed views over the water to the south and a grove of pines to the north (Fig. 14.4f). The mixed group of pines and red oaks in the middle of the park visually divides the east and west regions.

Finally, based on the existing parking place on the east side, we designed the trail from the entrance by placing markers and computing least cost path between these waypoints (Fig. 14.4e). The beginning point of the trail was the entrance and the end point was the shelter. In order to take the slope along the trail into consideration, we projected the profile of the trail below the model (Fig. 14.4e). The trail is then rendered in Blender as a boardwalk (Fig. 14.4h).

14.2.3   Scenario 2

After the first scenario, we restored the landscape to its original form. For the next design we decided to preserve most of the original topography, but create two smaller lakes (Fig. 14.5a). We built the shelter on top of a new mound so that it would have a prominent location with commanding views of the lakes (Fig. 14.5b). In contrast with previous scenario, we defined the park entrance at the corner of Centennial Parkway and Blair Road. The park trail connects the new entrance to the shelter. We routed it around the lakes to offer nice views, while reducing the slope along the trail.

In order to maintain open the views of the lakes from the shelter, we planted low shrubs in the middle of the park. We planted a colorful mix of species including red oaks and pines around the borders of the park to attract visitors from outside (Fig. 14.5f) and make the views from the shelter richer and more diverse.

Figure 14.5g shows the final result rendered in an abstract way, using low-poly visualization (see Sect. 5.7.1 for more details). We then used a head-mounted display to evaluate the 360° view from the shelter (Fig. 14.5h).

14.2.4   Evaluation of Scenarios

With Tangible Landscape we were able to rapidly develop multiple scenarios for a park—the first focused on a single large lake, the second on the journey through the landscape. With scenarios we can compare and contrast different ideas, solicit feedback from other designers, stakeholders, and the general public, and continue to refine our design. We compared and critiqued our two scenarios using the metrics, site plans, and perspective views shown in Fig. 14.6. With the metrics (Fig. 14.6a,b) and trail profiles (Fig. 14.6c,d) we can quantitatively compare the designs. The first scenario had significantly more water, while the second had gentler slopes along the trail and a more diverse landscape structure with more patches, more complex shapes, and more diverse distribution of plant species. The profile shows that the

Fig. 14.5 In the second scenario designers (**a**) sculpted the landscape to create two ponds and an artificial mound. Next, they (**b**) sited the shelter on top of the mound and designed a trail to connect it to the park entrance. Then, they (**c**) planted trees, (**d**) reviewed planting from a birds-eye view, and (**e**) explored various (**f**) realistic and (**g**) abstract views rendered both on the screen and (**h**) in the head-mounted display

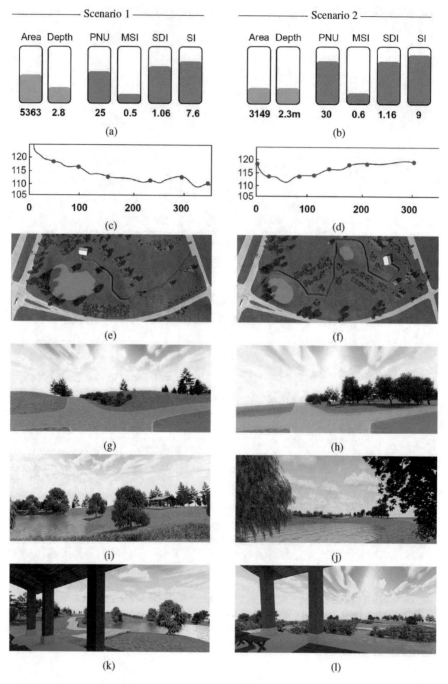

Fig. 14.6 A comparison of two design scenarios in terms of: (**a**),(**b**) hydrological and landscape metrics; (**c**),(**d**) trail profile; (**e**),(**f**) site plan (North up, scale ~1/10,000 m); and rendered views (**g**),(**h**) from the park entrance; (**i**),(**j**) from the lakes towards the shelter; and (**k**),(**l**) from the shelter to the lakes. *PNU* number of patches, *MSI* mean shape index, *SDI* Shannon diversity index, *SI* shape index

trails in both designs are too steep and need to be regraded at the entrance to the park. The trail in first scenario also has too steep a slope near the shelter. With the perspective views (Fig. 14.6e–l) we can subjectively judge the aesthetics of the scenarios. The first scenario used undulating landforms interspersed with trees and shrubs as a border for the park (Fig. 14.6g) and used specimen trees for visual interest within the park (Fig. 14.6i). The second scenario was designed to enclose the park with trees (Fig. 14.6h) and frame the wide, open views within (Fig. 14.6j). The first scenario has more interesting topography and picturesque views with specimen trees, while the second scenario is more walkable and ecologically diverse.

References

Baker, W. L., & Cai, Y. (1992). The r.le programs for multiscale analysis of landscape structure using the GRASS geographical information system. *Landscape Ecology, 7*(4), 291–302.

Molnar, D. (2015). *Anatomy of a park: essentials of recreation area planning and design.* Long Grove, IL: Waveland Press.

Appendix A

A.1 Applications of Tangible Landscape

In this section we briefly describe additional applications of Tangible Landscape that we, our students, and our colleagues have developed. For each project we explain the aim, used methods and studied processes.

A.1.1 Modeling Avalanches in High Tatras

Eva Stopkova modeled snow avalanches in the High Tatras mountain range in Slovakia (Stopková 2007, 2008). Avalanche potential is closely related to landscape topography so Stopkova built a physical model of her study area in High Tatras in order to visualize the spatial pattern of topographic parameters and landforms (Fig. A.1). She was able to project the results of avalanche danger assessment over the model, but she was not able usefully compute the assessment using the modified and scanned physical model. The scanned models were not accurate enough for practical applications due to their small scale and the steepness of their slopes. Modeling avalanches with Tangible Landscape would require a larger physical model at a larger map scale. CNC routing can be used to precisely and efficiently fabricate the size of model needed for such an application.

A.1.2 Visualizing the Evolution of Oregon Inlet

Liliana Velasquez Montoya created a projection augmented sand model of Oregon Inlet where she was modeling how storms drive the morphological evolution of barrier islands. To visualize and present the results of her research Montoya

© The Author(s) 2018
A. Petrasova et al., *Tangible Modeling with Open Source GIS*,
https://doi.org/10.1007/978-3-319-89303-7

Fig. A.1 A sand model of High Tatras in Slovakia augmented with a projection of the landforms computed by the module *r.geomorphon*; a legend and a diagram of the landform types are projected on the table beside the model

Fig. A.2 A sand model of Oregon Inlet with a projected video showing its formation and evolution along with a graph of wave height changing over time

projected a video depicting the evolution of the inlet in response to storm events onto a sand model. She built the sand model to physically represent the present 3D form of the inlet. Since the sand model was not dynamic it only matched the end of animation. This was a useful mode of representation as the model acted as a datum that helped us compare the present and past morphology of the inlet. Since the small sand model was only large enough to show the most active part of the inlet she also projected the surrounding coastal landscape (Fig. A.2).

A.1.3 Designing Disaster Relief Housing for Rodanthe

Logan Free designed disaster relief shelters for an emergency ferry terminal at Rodanthe on the Outer Banks of North Carolina. The ferry terminal is used for evacuation and relief when there are disasters like hurricanes or bridge failures. Free was tasked with designing temporary housing structures and a community service center for disaster recovery at the ferry terminal. To identify vulnerable hotspots and select safe sites for the shelters he built a sand model of the site and used Tangible Landscape with the module *r.lake* to simulate the potential extent of storm surge flooding (Fig. A.3).

A.1.4   Simulating Landscape Change in Charlotte

Douglas A. Shoemaker created a model of a small watershed in Charlotte, North Carolina to test how the FUTure Urban-Regional Environment Simulation (FUTURES) framework could be integrated with Tangible Landscape. FUTURES uses a stochastic patch growing algorithm to simulate urban growth and land use change (Meentemeyer et al. 2013). It has been integrated with GRASS GIS as the add-on module *r.futures* (Petrasova et al. 2015). Shoemaker used the raster digitizer tool in the GRASS GIS GUI to designate a site for development and change the land cover. He also sculpted the sand to grade the topography for the new development. Then he ran FUTURES to explore how the new development would influence longterm landscape change in the watershed-scale study area (Fig. A.4).

(a)   (b)

Fig. A.3 Simulating storm surge at the Rodanthe emergency ferry terminal on the Outer Banks of North Carolina: (**a**) the mean high water level on the sound side of the island and (**b**) the approximate water level during high storm surge

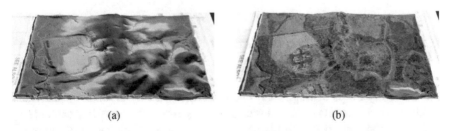

(a)   (b)

Fig. A.4 Landscape change modeling for a watershed in Charlotte, North Carolina: (**a**) a sand model of the terrain at the Charlotte study area augmented with a projected shaded relief to enhance perception of relief; (**b**) using a mouse and the digitizer tool to draw a polygon to designate a development area

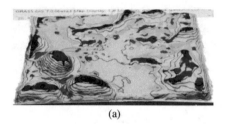
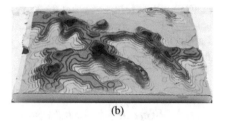

(a)   (b)

Fig. A.5 Reconstructing a paleolake and a terrain model used for cell tower planning. (**a**) A reconstruction of a paleolake in the Galuut Valley, Mongolia showing the landforms computed by *r.geomorphon*. (**b**) A physical model used to site cell towers in Athens county, Ohio

A.1.5 Reconstructing a Paleolake in Mongolia

Gantulga Bayasgalan used a sand model to reconstruct a paleolake in the Galuut Valley in central southwest Mongolia. He modeled alternative reconstructions of the landscape in sand and used Tangible Landscape to compute the flow accumulation pattern, geomorphological features, and several other characteristics for each landscape configuration (Fig. A.5a).

A.1.6 Cell Tower Planning in Athens County

Jesse Boyd used Tangible Landscape to site cell towers in Athens county, Ohio. To find good sites for cell towers, Boyd tested different spatial configurations by placing markers representing cell towers on a physical model of the landscape (Fig. A.5b). Since topography blocks cellular signals he used viewshed analysis to estimate cellular coverage. He experimented with different cell tower heights by changing the observer elevation parameter of the module *r.viewshed*.

A.1.7 Monitoring Coastal Erosion

Tristan Dyer used terrestrial lidar data to monitor coastal erosion on a beach near the US Army Corps of Engineers' Field Research Facility in Duck on the Outer Banks of North Carolina. Coastal erosion can be visually and tactilely explored by building a sand model for each lidar survey. Dyer scanned a transect of the beach with a terrestrial lidar system, interpolated the lidar point cloud as a DEM, and then built a sand model with the aid of Tangible Landscape to quickly prototype and visualize the landform (Fig. A.6a). A sequence of these models would show the morphological evolution of the beach.

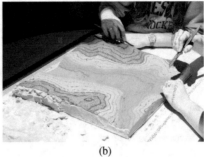

(a)   (b)

Fig. A.6 Sand models used to study beaches and streams. (**a**) A sand model of a beach in Duck, North Carolina used to study beach erosion. (**b**) A sand model of a valley in North Carolina Piedmont used to study how beaver dams influence stream flow and morphology

A.1.8 Exploring Impacts of a Beaver Dam

Karl W. Wegmann and his students used Tangible Landscape to study how beaver dams affect streams in North Carolina. They used Tangible Landscape to model the morphology and flow of a stream with and without beaver dams. First they built a sand model of a segment of the existing stream with beaver dams and studied its topographical parameters and hydrological processes. Then they sculpted the sand model and computed different water flow scenarios to study how the stream could have been without beaver dams (Fig. A.6b).

In this application we merged the scanned elevation model with a DEM of the surrounding landscape to put the modeled area into its broader context and account for water flowing from the entire contributing watershed. The water flow simulation was then run over the entire area so that the water from the contributing watersheds would fill the stream.

A.1.9 Modeling the Potential Impacts of a Coal Ash Pond Spill

Matthew Horvath (2014, Spill impacts from coal ash pond using GRASS GIS, Unpublished poster) studied the potential impacts of spills from the Cape Fear Plant coal ash ponds near Moncure, North Carolina. Horvath built a sand model for a study area near the Cape Fear river and used Tangible Landscape with the add-on module *r.damflood* to simulate a breach in a coal ash pond (Fig. A.7). Artificial landforms kept the simulated spread of contaminated surface water from reaching the nearby Cape Fear river.

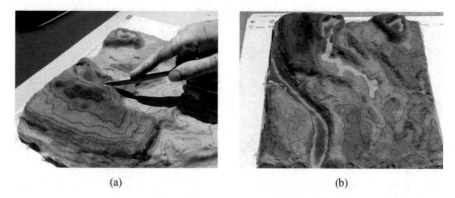

(a)   (b)

Fig. A.7 Making a breach in a coal ash pond (**a**) and simulating the spread of contaminated water (**b**)

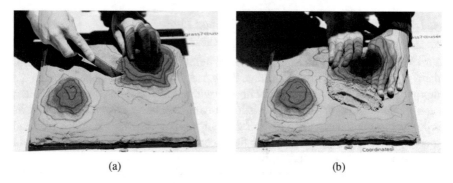

(a)   (b)

Fig. A.8 Iteratively sculpting a landform in sand to generate time-series data: (**a**) the initial state followed by (**b**) changing the position and shape of the landform

A.1.10   Testing a Landform Migration Algorithm

Petras et al. (2015) used Tangible Landscape to test the behavior of a newly developed algorithm for describing landform migration. The algorithm maps the horizontal migration of complex landforms as gradient fields based on the analysis of contour evolution. The gradient field is a vector field representing the movement and deformation of contours. It can be used to quantify the rate and direction of landform migration at any point in space and time. Data for testing algorithms is usually computer generated. In this case, however we wanted to create simplified, yet realistic abstractions of landforms that would allow us to clearly model and visualize landform migration. It was easier to make these abstract landforms by hand than on the computer. To create test data in a controlled environment we built sand models by hand to represent a moving landform. Using Tangible Landscape we sculpted and scanned a model of the initial terrain and then sculpted and scanned a model of the terrain after it had migrated to create an elevation time-series (Fig. A.8).

(a)   (b)

Fig. A.9 Using object recognition to digitize polygons and pixels representing treatment areas in order to computationally steer (**a**) Sudden Oak Death in Sonoma Valley, California, and (**b**) termites spread simulation in Fort Lauderdale, Florida

A.1.11   Managing the Spread of Sudden Oak Death in Sonoma Valley

In collaboration with Francesco Tonini and Douglas A. Shoemaker we used Tangible Landscape to test the effectiveness of different treatment scenarios for controlling the spread of Sudden Oak Death in Sonoma Valley, California (Tonini et al. 2017). We used a stochastic model to simulate the spread of Sudden Oak Death and Tangible Landscape to visualize the spread and designate new treatment areas. We CNC routed a large (1 m × 1 m) MDF model of the terrain and coated it with magnetic paint. To make markers that would hold in place on the complex topography we attached adhesive magnetic strips to the base of basswood markers. We used the markers to collaboratively draw polygons representing treatment areas with Tangible Landscape (Fig. A.9a). The polygons reduced the host species of the disease on aligned pixels thus influencing the next run of the simulation. Each treatment area was automatically assigned a label with its area. The sum of all of the treatment areas was dynamically displayed on a simple dashboard. The disease spread simulation was written in the R (R Core Team 2013) statistical programing language and environment.

A.1.12   Participatory Modeling Workshop for Managing Sudden Oak Death in Oregon

In collaboration with Devon A. Gaydos, Richard C. Cobb, and Ross K. Meentemeyer leveraging experience from the previous Sudden Oak Death work (Tonini et al. 2017) we held a participatory modeling workshop at Oregon Department of Forestry in Salem, Oregon with study area near the coast in Curry County, Oregon. We again used the aforementioned stochastic model to simulate the spread of Sudden Oak Death disease. Tangible Landscape was the tool to visualize the disease spread and designate new treatment areas. We CNC routed six tiles creating a 120 cm × 80 cm model. The tiles were from high density foam and hollow to make them light for traveling. The model used custom projection which was rotated 90° to accommodate rectangular shape of Kinect's field of view and the study area elongated in north-south direction. We used red felt to represent treatment areas which were then used to modify the tree density inputs for the spread model. We used two separate USB buttons next to the physical model to run the spread model and to view different stochastic results. The results were pushed to a web-based dashboard which was recording global statistics for each run. An additional desktop dashboard was providing quick feedback about the cost of currently proposed treatment. More than ten participants worked with us designing and experimenting with various spatial combinations of treatments while cutting and combining the pieces of felt.

A.1.13 Managing the Spread of Termites in Fort Lauderdale

In collaboration with Francesco Tonini we used Tangible Landscape as a spatial decision support tool for collaboratively managing the spread of termites in Dania Beach, Fort Lauderdale, Florida. We used pixel-based object recognition to interact with a stochastic simulation of the spread of termites. Placing a marker on a flat surface with a projected grid changed a value on the corresponding pixel in the simulation input—a raster index of termite habitat (Fig. A.9b).

A.1.14 Tangible Exploration of Subsurface Data

Using Tangible Landscape we explored the 3D distribution of the percentage of subsurface moisture (Petrasova et al. 2014) measured in Kinston, North Carolina and represented as a GRASS GIS 3D raster (GRASS Development Team 2015). By digging into the sand model with our hands we could explore underground as if we were at an excavation site. As we dug into the sand the cross-section of the scanned surface with the 3D raster representing soil properties was projected in real-time (Fig. A.10). We also projected additional GIS layers (such as an orthophotograph, elevation contours, and flow accumulation) over the model to provide spatial context and gain further insights into soil moisture distribution and its relation to the landscape.

Additionally, in combination with marker detection from Sect. 4.3 we can create and visualize vertical profiles or soil core samples. This method is an intuitive and natural way of exploring subsurface data and it represents an alternative to more abstract 3D computer visualization tools.

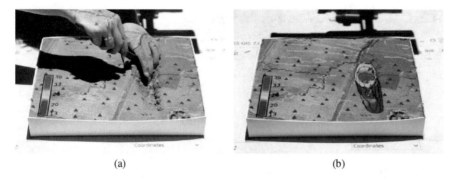

(a)   (b)

Fig. A.10 Exploring subsurface moisture by (**a**) removing sand and (**b**) projecting the cross-section of a scanned surface with the 3D raster of moisture

A.2   Data Sources

Throughout the book we use various data sources. Here is a list of the most useful data sources for the study areas presented in the book.

A.2.1   Sample Data for This Book

NCSU GeoForAll Lab

http://geospatial.ncsu.edu/osgeorel/data.html

A.2.2   US Lidar Data

United States Interagency Elevation Inventory
http://coast.noaa.gov/inventory/
Earth Explorer
http://earthexplorer.usgs.gov/
Digital Coast
http://coast.noaa.gov/dataviewer/
Open Topography
http://www.opentopography.org/

A.2.3   US Digital Elevation Models

National Elevation Dataset on the National Map Viewer
https://viewer.nationalmap.gov/basic/

A.2.4   US Orthoimagery

USGS NAIP Orthoimagery web mapping service
https://services.nationalmap.gov/arcgis/services/USGSNAIPPlus/MapServer/
WMSServer
USGS NAIP Orthoimagery on the USGS Earth Explorer
https://earthexplorer.usgs.gov/

A.2.5   US Soil Surveys

USDA Web Soil Survey
http://websoilsurvey.sc.egov.usda.gov

A.2.6   US Fire Modeling Data

LANDFIRE
http://www.landfire.gov/

A.2.7   Global Datasets

GRASS GIS community list of global datasets
http://grasswiki.osgeo.org/wiki/Global_datasets

A.2.8   NC Climate Data

State Climate Office of North Carolina
http://www.nc-climate.ncsu.edu/

Fig. A.11 Starting GRASS GIS

A.3 Starting with GRASS GIS

To run examples from this book we provide a dataset which can be downloaded from
the NCSU GeoForAll Lab website.[1] Once the downloaded dataset is unzipped we
create a directory grassdata in our home directory and place the dataset there.
Then we start GRASS GIS and a start up dialog in Fig. A.11 appears on screen.
In the upper text field we specify the path to our grassdata directory. Then we
should be able to see Location nc_spm_tl containing several Mapsets. We select
Mapset practice1 and press the Start button.

After opening the Layer Manager and Map Display windows we display the
raster map elevation as shown in Fig. A.12. To execute commands from this
book presented in boxes with green background we paste the commands into the
Console tab as shown in Fig. A.13a and press Enter. The Python code in boxes with
blue background can be executed in the *Python* tab (Fig. A.13b) or in any Python
shell running in a GRASS GIS session.

---

[1]http://geospatial.ncsu.edu/osgeorel/data.html.

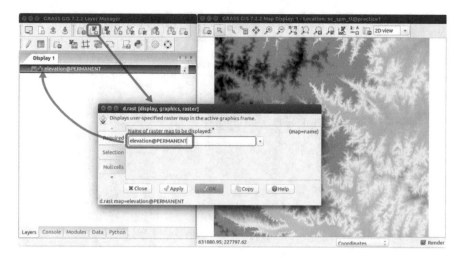

Fig. A.12 Display raster map elevation: click on *Add raster map layer* button in toolbar, select `elevation` from drop-down list, and press OK

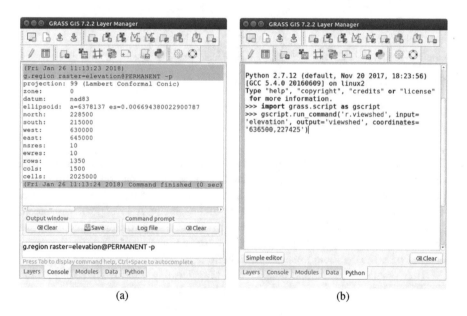

Fig. A.13 GRASS GIS modules can be run from (**a**) the *Console* tab using command line syntax or from the (**b**) *Python* tab using Python syntax

More instructions how to use GRASS GIS GUI are available on the official GRASS GIS introduction page[2] and in the GRASS GIS GUI manual.[3]

References

GRASS Development Team (2015). 3D raster data in GRASS GIS. Retrieved August 13, 2015, from http://grass.osgeo.org/grass70/manuals/raster3dintro.html.

Meentemeyer, R. K., Tang, W., Dorning, M. A., Vogler, J. B., Cunniffe, N. J., & Shoemaker, D. A. (2013). FUTURES: multilevel simulations of emerging urban-rural landscape structure using a stochastic patch-growing algorithm. *Annals of the Association of American Geographers*, *103*(4), 785–807.

Petras, V., Mitasova, H., & Petrasova, A. (2015). *Mapping gradient fields of landform migration* (pp. 173–176). Poznań, Poland: Bogucki Wydawnictwo Naukowe, Adam Mickiewicz University in Poznań - Institute of Geoecology and Geoinformation.

Petrasova, A., Harmon, B., Mitasova, H., & White, J. (2014). Tangible exploration of subsurface data. In *Poster presented at 2014 Fall Meeting*, 15–19 Dec. San Francisco, CA: AGU.

Petrasova, A., Petras, V., Shoemaker, D. A., Dorning, M. A., & Meentemeyer, R. K. (2015). The integration of land change modeling framework FUTURES into GRASS GIS 7. In *Geomatics Workbooks n 12 – "FOSS4G Europe Como 2015"*.

R Core Team (2013). *R: A language and environment for statistical computing*. Vienna, Austria: R Foundation for Statistical Computing. http://www.R-project.org/.

Stopková, E. (2007). Modelovanie oblastí vzniku lavín s využitím GIS. In *Sborník studentské konference GISáček*.

Stopková, E. (2008). *Predikcia lavínového nebezpečenstva s využitím GIS*. Bachelor's thesis, Slovak University of Technology in Bratislava.

Tonini, F., Shoemaker, D., Petrasova, A., Harmon, B., Petras, V., Cobb, R. C., et al. (2017). Tangible geospatial modeling for collaborative solutions to invasive species management. *Environmental Modelling & Software*, *92*, 176–188.

---

[2]http://grass.osgeo.org/grass72/manuals/helptext.html.

[3]http://grass.osgeo.org/grass72/manuals/wxGUI.html.

Index

© The Author(s) 2018
A. Petrasova et al., *Tangible Modeling with Open Source GIS*,
https://doi.org/10.1007/978-3-319-89303-7

Printed in the United States
By Bookmasters